Systems Thinking

Selected Readings

Edited by F. E. Emery

Penguin Education

Penguin Education
Penguin Books Ltd, Harmondsworth,
Middlesex, England
Penguin Books Inc., 7110 Ambassador Road,
Baltimore, Md 21207, U.S.A.
Penguin Books Australia Ltd, Ringwood,
Victoria, Australia
Penguin Books Canada Ltd,
41 Steelcase Road West, Markham, Ontario, Canada
Penguin Books (N.Z.) Ltd,
182–190 Wairau Road, Auckland 10, New Zealand

First published 1969
Reprinted 1970, 1971, 1972, 1974, 1976

Made and printed in Great Britain by
Richard Clay (The Chaucer Press) Ltd,
Bungay, Suffolk
Set in Monotype Times

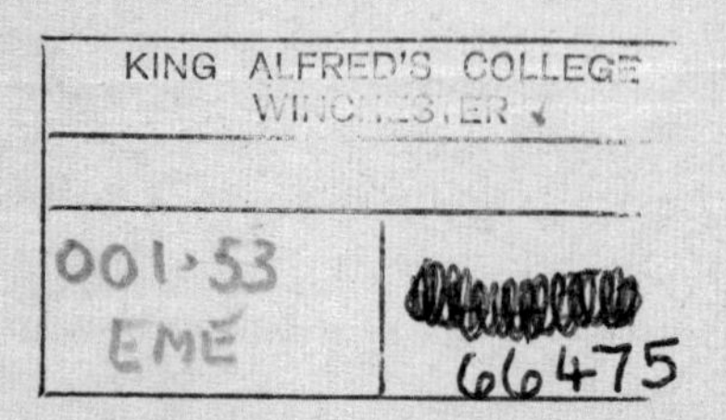

Contents

Introduction

In the selection of papers for this volume, two problems have arisen, namely what constitutes 'systems thinking' and what systems thinking is relevant to the thinking required for organizational management? The first problem is obviously critical. Unless there were a meaningful answer, there would be no sense in producing a volume of readings in systems thinking in any subject. A great many writers have manifestly believed that there is a way of considering phenomena which is sufficiently different from the well-established modes of scientific analysis to deserve the particular title of systems thinking. Reasons for believing that the distinction is of value are spelt out in the first selections.

There have in fact been two arguments for a systems approach to the analysis of living phenomena.[1] First has been the argument that only such an approach will reveal the 'Gestalten' properties that characterize the higher levels of organization which we call 'living systems'. It will be noticed that we are parenthesizing the key terms. This pretty well indicates the uncertain state of knowledge in this field. Second has been the argument that many of these Gestalten properties are common to the different levels of organization of living matter (from bacteria to human societies) and hence provide a valid and powerful form of generalization. There is at least one further line of argument although it has had little apparent attraction to the main contributors to the systems approach. This is that a systems analysis of living organizations is likely to reveal the 'general in the particular'. Analysis of part systems in cause–effect terms, for example, of liver disorders, death rates, recruitment, training, or productive efficiency, builds up a certain kind of knowledge. However, the total systems of

1. Throughout the volume we have kept to the strand of thought that runs from theorizing about biological systems in general to social systems. We have practically ignored the strand that arises from the design of complex engineering systems. Through such movements as operations research and cost–benefit analysis this influence is being strongly felt by management but its methods and language are so different as to require separate treatment.

which they are a part usually offer alternative paths which will minimally meet organizational requirements and/or provide substitute feedback control systems. Analysis of the total system is likely to reveal those properties, general to the species, that have enabled the species to adapt and survive in its typical environment. This line of argument clearly accepts the first argument, namely that there are Gestalten qualities of living organizations that are unlikely to be revealed by the ordinary modes of scientific analysis. It goes beyond this in suggesting that there may be properties that can be generalized to the 'species' and yet claim no necessary generalizability to all living systems because systems analysis presupposes a knowledge of what functions the part systems can undertake. Without this knowledge it would be very difficult indeed to determine what the total system was supposed to be co-ordinating and controlling.

We posed a second question, 'What systems thinking is relevant to the thinking required for organizational management?' The editor believes that it has been shown that living systems, whether individuals or populations, have to be analysed as 'open systems', i.e. as open to matter–energy exchanges with an environment. Human organizations are living systems and should be analysed accordingly. The fact that it faces us with the task of analysing forbiddingly complex environmental interactions gives us no more of an excuse to isolate organizations conceptually than the proverbial drunk had when searching for his lost watch under the street lamp because there was plenty of light when he knew he had lost it in the dark alley.

A great deal of work has been done in the social sciences to elucidate the properties of social systems considered as isolated entities. We shall not draw on this material. Management is concerned with the control of social systems, technologies, *and* markets. Our central purpose in selection has been, therefore, to depict the emergence and clarification of the view that living systems are essentially 'open systems', not 'closed systems'. Despite Koehler's contribution (Reading 3) in 1938 and Angyal's (Reading 1) in 1941, the major impact came from von Bertalanffy (Reading 4) in 1950. It is perhaps unfortunate that this impact gave rise to a movement for General Systems Theory along with its search for dynamic principles common to all kinds of systems living or mechanical. This movement has been attractive to those with a systems engineering orientation, but has so far failed to further its unifying mission and has tended to overshadow the early recognition by Ashby and Sommerhoff that if living systems are to be treated as open systems we must be able to characterize their environments.

In Part Three we present a set of papers which have sought to deal specifically with the properties of environments that are relevant to adaptive behaviour.

These efforts are all concerned with global properties of organizational environments. In this capacity these theories seem to be far removed from the specificity of organization–environment relation that Sommerhoff shows to be necessary for a science of living systems. The gap may well prove to be more apparent than real. As Gibson (1966) and Tomkins (1953) have argued, living systems probably learn and hence adapt because of their ability to react to the general and less variable properties of the environment, rather than because of their sensitivity to the concrete events and objects which do after all yield a constant flux of stimulation. Measuring organizational environments along the dimensions suggested by Simon, Ashby, and the others, may be all that is required to realize Sommerhoff's general theory.

Part Four brings together papers that concern the extension of these notions to social systems. It will be clear that these efforts have barely begun to encompass the richness of thought exhibited in the preceding sections. There is no reason why the serious student of management should not regard this as a challenge to join in the bridging operation. To encourage this we present six principles which reach back as far as the work of Koehler in order to identify lessons of value to systems management:

1. The primary task of management is to manage the boundary conditions of the enterprise. The boundaries of an enterprise are those levels of exchange with the environment which allow it to survive and grow. They can be managed only by managing the co-variation of internal and external processes. In so far as a manager has to co-ordinate or otherwise resolve internal variances then he is distracted from his task.

2. The goals or purposes of an enterprise can be understood only as

special forms of interdependence between an enterprise and its environment. They cannot be identified with the state of equilibrium that is the *end-state* for closed systems. The state of equilibrium represents a minimum level of potential energy or capacity for work. *The enterprise seeks to establish and maintain those forms of interdependence that enable it to maximize its potential energy or capacity for work.* As in Koehler's example of the flame (pp. 63–5), a steady state is achieved only at the level of maximum potential energy. The form of this potential capacity and the exchanges are determined by the special forms of interdependence into which the enterprise enters, but achievement of a steady state is the most general dynamic trend in an open system.

3. An enterprise can achieve a 'steady state' only when there is (a) constancy of direction, i.e. despite changes in the environment or in the enterprise, the same outcomes or focal conditions are achieved. Put another way, the system remains oriented to the same end; (b) that with respect to that end, the system maintains a *rate of progress* toward it which is within limits defined as tolerable. A more precise statement of 'rate of progress' might be that the enterprise achieves the required focal condition with lesser effort, with greater precision for relatively no more effort, or under conditions of greater variability. In any of these cases, the level of exchange would be more favourable to the enterprise. One implication of this proposition is that an enterprise can have no equilibrium state such as can be found in physical systems (because in the former case, the relevant internal and external variables are capable of independent variation – the state of one does not automatically determine the other). A positive implication is that an enterprise cannot hope to achieve steady state (except accidentally) unless it sets a mission for itself in terms of outcomes that are capable of achievement and yet are sufficiently beyond present performance to allow for some measurable degree of progress.

4. Given the last two propositions, *the task of management is governed by the need to match constantly the actual and potential capacities* of the enterprise to the actual and potential requirements of the environment. Only in this way can a mission be defined that may enable an enterprise to achieve a steady state. However, the actions of management cannot in themselves constitute a logically sufficient condition for achievement of a steady state.

5. A 'steady state' for the system cannot be achieved by any finite combination of regulatory devices or mechanisms that are aimed at achieving a steady state for some partial aspect of the system such as input–output rates, internal change, or environmental contact. *In a human organization, the two requirements for a steady state, unidirectionality and progress, can be achieved only by leadership and commitment.*

The end-state of the system must be clearly enough defined and agreed upon to enable the system to be oriented toward it regardless of a wide range of changes in their relations. Secondly, the members of the organization must be so committed to the end-state that they will respond to emergencies calling for greater efforts. The basic regulation of open systems is thus self-regulation – regulation that arises from the nature of the constituent parts of the system.

One corollary is that it is only within this framework that regulatory mechanisms, such as cost controls, can make an effective contribution. In creating these mechanisms it is essential to ensure that they do not run counter to, or undermine the requirements for self-regulation, and to remember that mechanisms which are appropriate in one phase of a system's existence may, with a change in location with respect to the mission, become inappropriate.

The measure of whether these processes of self-regulation are operating effectively, of whether the system is healthy and maturing, is to be found in the steady growth of potential capability with respect to the mission. In the case of any enterprise, the critical question at any time is whether it is more capable than before of fulfilling the tasks arising from its mission. A good record of recent performance, e.g. high profit yield, would not in itself exclude the possibility that potential capacity had in fact been reduced.

6. An enterprise can only achieve the conditions for a steady state if it allows to its human members a measure of autonomy and selective interdependence. This proposition is clear enough when applied to organizations composed of professionals. It is less clear that it applies to enterprises in general because it introduces an assumption which is new in this context, namely that individuals themselves have open-system characteristics and can be related to each other or to organizations only in ways that are appropriate to such systems. In particular, commitment presupposes that the individual has sufficient autonomy to exercise choice. The requirement that the co-ordination of components be maximally brought about by themselves (proposition five) requires some sacrifice of autonomy and to that extent threatens commitment. This threat can be lessened by allowing selective interdependence.

These principles barely draw upon the potential of systems theory for management theory. They also show little evidence of a coherent and comprehensive theory from which such principles could be rigorously deduced. Nevertheless it is in finding new ways of looking at things that men have managed in the past to advance scientific understanding.

We have tried to select readings that would excite readers to a new way of looking at mundane realities of human organization. They have been arranged to lead from the earlier to the later papers and from consideration of systems in general to social systems in particular.

For Part One we have selected the earlier papers that explicitly or implicitly argue for a new logic in the study of complex systems that display purposive or adaptive behaviour. What these authors have in mind is something more than the model of causal analysis usually associated with the physical sciences; it is also something less than a rigorous predictive theory of some restricted area of behaviour. This latter feature has occasioned some misguided criticism to the effect that theories that cannot predict and hence cannot be experimentally confirmed or disconfirmed are not scientific theories. Such criticism overlooks the all-important role in scientific development of our guiding metaphors and our principles for mapping the real world (Kuhn, 1962). If systems theorizing improves the 'internal coherence, implicative structure, freedom from the clutter and comprehensiveness' (Chein, 1967) of the maps we make of human organizations, then we must judge it as a scientific advance.

The selections in Part One only partly argue their case by pointing to principles of system action that can be inferred from their definition of a system. However, the papers in the remaining sections were specifically selected to sample the variety of efforts that have been made to meet scientific requirements of coherence, mplicative structure, etc. These efforts have been very varied. It is almost as if the pioneers, while respectfully noting each other's existence, have felt it incumbent upon themselves to work out their intuitions in their own language, for fear of what might be lost in trying to work through the language of another. Whatever the reason, the results seem to justify this stand-offishness. In a short space of time there has been a considerable accumulation of insights into system dynamics that are readily translatable into the differing languages and with, as yet, little sign of the divisive schools of thought that for instance marred psychology during the 1920s and 1930s. Perhaps this might still happen if some influential group of scholars prematurely decide that the time has come for a common conceptual framework.

To give ready access to this variety we decided to select in terms of individual theorists instead of conceptual areas such as co-ordination, control, regulation, temporal integration. Likewise, to allow space for each theorist to develop his body of concepts and principles we have had to restrict the range of the sample and to leave out articles of great merit as well as those which are essentially polished rehashes or reviews. We can only hope that the heaviness of the resulting selection will be found justified by the number of unmined diamonds that exists in even the earliest.

The importance of a solid ground in this new and explosive field is such that we have not tried to depict the frontiers of systems theorizing, except in the last section introducing notions of systems management. For those wishing to extend their range of coverage in the areas represented here they could do no better than turn to the *General Systems Yearbooks*. For those eager to see what is happening at the frontiers the best single source is probably the journal *Artorga*.

The editor is grateful to the Center for Advance Studies in the Behavioral Sciences, Stanford, for the peace and quiet he needed to prepare these Readings.

References

CHEIN, I. (1967), 'Versity vs. truth in the scientific enterprise', address to *Division of Philosophical Psychology, American Psychological Association*, 3 September.

GIBSON, J. J. (1966), *The Senses Considered as Perceptual Systems*, Houghton Mifflin.

KUHN, T. (1962), *Structure of Scientific Revolutions*, University of Chicago Press.

TOMKINS, S. S. (1953), *Affect, Imagery and Consciousness*, Springer, vol. 1.

Part One **Precedents to Systems Theory**

Reading 1 by Angyal provides a forceful statement of the view that the key concepts used to describe the organization of living systems require a new logic. In his subsequent work (Angyal, 1965) he did much to justify his assertion. Both of Angyal's major works should be consulted. They are focussed upon the individual biosphere (man in his environment) but brilliantly expose the general dynamics of system integration, in time as well as in space, of system dysfunctioning, and the dynamics involved in system change.

It is as well to be reminded that many of the more valuable insights into systems functioning, particularly the earlier insights, do not come labelled as systems theory. Feibleman and Friend (Reading 2) wrote as philosophers but penetrated deeply into the conceptual problems of what we are now accustomed to call systems theory. They realized that the major concern in organizational analysis would be with logically correlative relations (see Sommerhoff's theory of directive correlations in Reading 6), and they sensed that in analysis of any one level of system organization it probably would be adequate to consider only the ones immediately above and below.

Only pressing problems of space precluded a selection from S. C. Pepper (1950). This is of particular importance because the 'root metaphors' he identifies and rigorously defines are all clearly operating in different systems theorists and account for much of the mutual incomprehension that exists among them. 'Contextualism' is the root metaphor which comes closest to our bias in selecting for this volume.

References

ANGYAL, A. (1965), *Neurosis and Treatment*, Wiley.
PEPPER, S. C. (1950), *World Hypotheses*, University of California.

1 A. Angyal

A Logic of Systems

Excerpt from chapter 8 of A. Angyal, *Foundations for a Science of Personality*, Harvard University Press, 1941, pp. 243–61.

The Structure of Wholes

The problem of the integration of part processes in the total organism is the most important and at the same time the most difficult problem for a science of personality. The difficulty lies not alone in the paucity of usable factual data, but to an even greater extent in the inadequacy of our logical tools. Such a handicap is felt not only in the study of personality, but in the study of wholes in general. An attempt will be made to develop some concepts which may be useful for the understanding of the structure of wholes.

Our scientific thinking consists prevalently in the logical manipulation of relationships. That the structure of wholes cannot be described in terms of relationships has, however, been repeatedly pointed out by many writers. While accepting the premise that holistic connexions cannot be resolved into relationships, some authors have implied that the pattern or structure of wholes does not lend itself at all to logical manipulation. We suggest, however, that the structure of wholes is perhaps amenable to logical treatment after all, that, though it may not be described in terms of relations, it may be described in terms of some more adequate logical unit, representing an entirely different logical genus. Here the attempt will be made to demonstrate that there is a logical genus suitable to the treatment of wholes. We propose to call it *system.*

The ideal would be to develop a logic of system to such a degree of precision that it might offer the basis for exact mathematical formulation of holistic connexions. A. Meyer states, 'The mathematics which would be needed for the mathematical formulation of biological laws does not exist today. It has to be created by the new biology' (Meyer, 1934, p. 35). To construct a logic of systems which would be the counterpart of the conventional logic of relations is in itself a gigantic task and cannot even be attempted here. It may be true that a substantial advance in the study of wholes, and specifically in the study of personality, will come from the

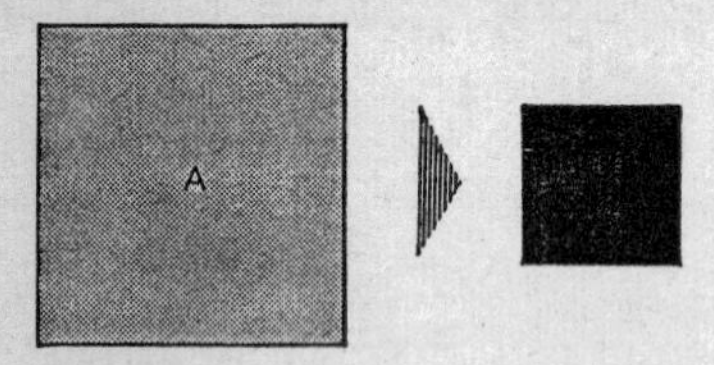

Figure 1

development of a logic of systems. We must, however, content ourselves for the present with the clarification of some aspects of the logical properties of systems and with the application of the insight thus gained to our specific subject-matter.

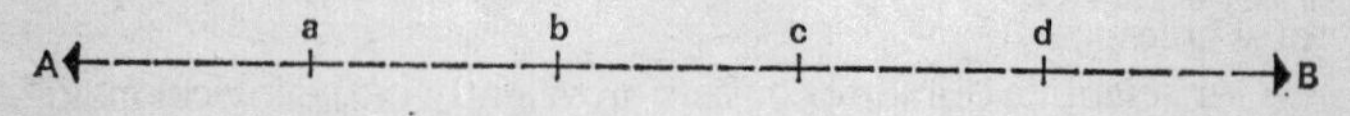

Figure 2

In order to demonstrate some of the logical characteristics of systems we may compare them with better known logical forms, namely, with relationships. As an example of a relationship we may take a quantitative one (Figure 1), and as an example of a system we may take a simple geometrical one (Figure 2) in which the points a, b, c, and d are parts of the simple geometrical system, the line A–B.

The differences between relationships and systems may be formulated as follows:

1. A relation requires two and only two members (relata) between which the relation is established. A complex relation can always be analysed into pairs of relata, while the system cannot be thus analysed. A system may involve an unspecified number of members. A system is not a complex relation. It is impossible to say what the relation between a and b, b and c, c and d, etc., should be in order to form a linear system.

I am aware of the fact that the restriction of the term 'relation' to cover only two-term connexions deviates from the contemporary usage of this term. Usually the term is employed also to include logical connexions that involve more than two terms. Such connexions, however, seem to fall into the following two categories:

(a) Compound relations, which can be reduced to two-term relations. One or both of the relata may be groups involving a more or less large number of members. Group A may include a, b, c, and d, and group B may include e, f, g, and h. When, however, group A is related to B, or to one member of B, the group is taken as a totality, that is, as *one* unit. A compound relation may involve also a chain connexion, for instance, a 'causal' sequence: a–b–c–d. It is clear that a compound relation can easily be resolved into two-term relations: a–b, b–c, c–d. A more complex relationship would be as in Figure 3.

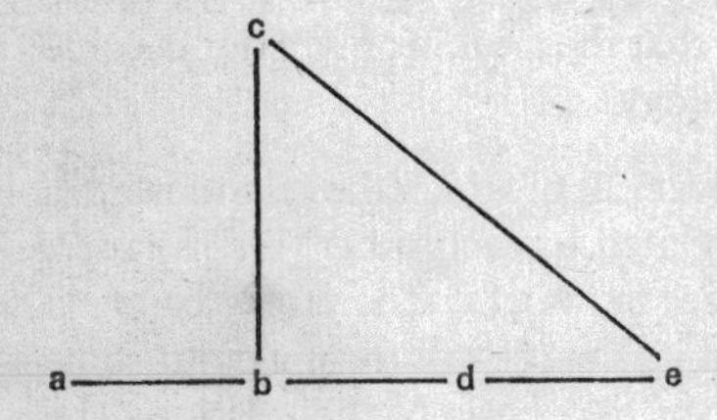

Figure 3

which consists of the two-term relations: a–b, b–c, b–d, d–e, c–e. The complexity may be even greater and yet a reduction to two-term relations still is possible.

(b) There are also connexions which involve more than two terms and cannot be reduced to two-term relations as, for instance, b is between a and c. It appears, however, that those compound connexions which cannot be reduced to two-term relations exhibit all those qualities which are set forth in this discussion as the characteristics of systems. One may, of course, use the term 'relation' in a very broad sense as is commonly done, but then one must admit that one subsumes two very different logical genera (two-term relationships and complex connexions which are reducible to two-term relations on the one hand, and complex connexions which cannot be reduced to two-term relations on the other hand) under the same term.

The term 'system', as used in this discussion, is also at variance with the common usage. Usually one designates by system any aggregate of elements considered together with the relationships holding among them. It will be shown in the following discussion that the type of connexions in a whole is very different from the connexions which exist in an aggregate. The term 'system' is used here to denote a *holistic system.* Further, in using this term we abstract *constituents* ('elements') and refer only to the *organization* of the whole. Thus 'system' for our discussion is holistic organization.

It might seem desirable in the present discussion to substitute other terms for 'relation' and 'system'; however, it seems to me equally desirable to avoid the coining of terms. Since I have pointed out the differences between my usage and the common usage of the terms I may hope that the argument will not become obscured through this terminology.

2. A relation requires an aspect out of which the relationship is formed. Two objects can be related to each other, for instance, with regard to their color, size, or weight. Therefore, before a relationship can be established it is necessary to single out some aspect of the relata which serves as a basis of the relation. The attribute of the relata on which the relationship is based is an immanent quality of the object, like size, color, or weight. The object enters into a relationship with another object because of its immanent qualities. Most relationships are based upon 'identity', diversity, or similarity (partial identity with partial diversity) of the object, that is, on immanent attributes. The members of a system, on the contrary, do not become constituents of the system by means of their immanent qualities, but by means of their distribution or arrangement within the system. The object does not participate in the system by an inherent quality but by its *positional value in the system.* It is immaterial for a linear system whether points or stars or crosses or circles or any other objects be the members, if only in the arrangement the positional values remain the same.

Between the constituents of a system, after they gain a positional value from the system, further relations may be established *in a secondary way* which are not based on the immanent properties of the relata, but on their secondary positional value. Such relationship is, for instance, A is below B. Such relations are secondary: it is presupposed that the members have a positional value in a system of coordinates.

3. In establishing a relationship between objects and in arranging objects in a system, the separation of the objects is presupposed. Multiplicity of objects is only possible in some kind of *dimensional domain* (a manifold). The clearest examples of dimensional domains are space and time, which have been re-organized by philosophers for a long time as *principia individuationis*, that is, domain which make possible a multiplicity of individual objects. We cannot speak of *two* objects unless they are placed in different points of time or in different points of space or unless a distance between them is not established in some other kind of dimensional domain.

Although the dimensional domain is a necessary condition for both relationships and systems, the function of the dimensional domain is different for these two logical genera. The role of the dimensional domain for a relationship is merely disjunction of the relata. But the role of the dimensional domain ends here. The domain itself does not enter into the relationship. For instance, if two colors are separated in space a comparison between them can be made without any further reference to space.

The dimensional domain is more intimately involved in the formation of systems. Here the dimensional domain not only separates the parts, but it participates in the formation of the system. The system is dimensional. *A system is a distribution of the members in a dimensional domain.*

4. In a relationship the connectedness between the relata is a *direct* one. The connexion goes without any mediation directly from a to b and vice versa. The connexion between the members of a system is, however, of a more complex type. Although there is a connexion between the points a, b, c, d when they form a straight line, this connexion is not a direct one in our sense. It is impossible to say what relationship should connect a with b, and

c with d, and a with d, etc., to form a linear arrangement. In this example the members of the system which are points are linearly connected only by forming a whole. System-connectedness of the parts cannot be expressed as a–b, b–c, a–c, but as in Figure 4.

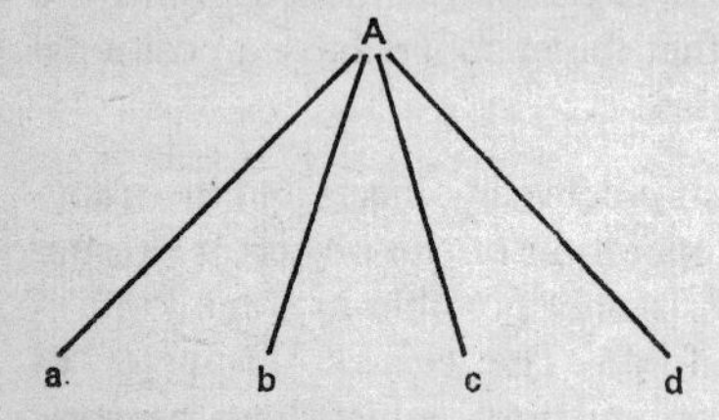

Figure 4

In a system the members are, from the holistic viewpoint, not significantly connected with each other except with reference to the whole.

The constituent parts of a system are not considered separately but with respect to a superordinate, more inclusive factor, the

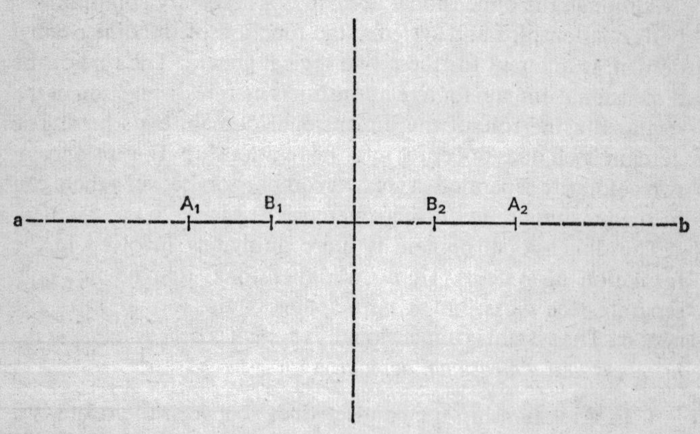

Figure 5

system in and by which they are connected. An interesting example of this state of affairs is given in the fact of geometrical symmetry. Such an arrangement involves two figures which are identical in shape and size, and show a special kind of correspondence with regard to their positions in space. The identity of geometrical figures can be demonstrated by bringing them together in space, in which case they must coincide. Symmetrical figures, however, can coincide only under special conditions. Taking one-dimensional geometrical figures such as those in Figure 5, it is clear that such figures cannot be made to coincide by moving them in a one-dimensional space, that is, along the line a–b. A_1 will not coincide with A_2 and B_1 with B_2 at the same

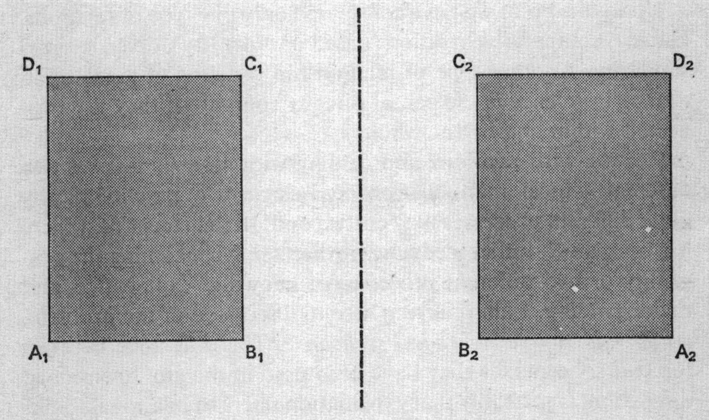

Figure 6

time. To make them actually coincide they must be rotated within a two-dimensional space with respect to the axis of symmetry. Thus the position of the two lines is analogous, not in a one- but in a two-dimensional space. Taking two-dimensional symmetrical figures such as those in Figure 6, it can easily be seen that they cannot be made to coincide in any way in the concrete if one shifts them within a two-dimensional space, but only within a three-dimensional space. Three-dimensional symmetrical objects again cannot be made to coincide in a three-dimensional space. That would require in the concrete a fourth dimension. If one's mirror image were real and one could go beyond the mirror, one still could not step into one's mirror image and completely coincide with it, because the right side of the person would be on the left side of the mirror image and vice versa; a right-hand glove cannot be made into a left-hand glove; and so on. For the coincidence of symmetrical figures, a space which has one dimension more than the number of dimensions of the symmetrical figures is required. Thus the congruence of one-dimensional symmetrical figures is brought about in a two-dimensional space, and that of two-dimensional figures in a three-dimensional space. Symmetry seems to be psychologically a kind of system-connectedness, that is, a connectedness of parts, not between themselves but in a superordinate whole.

Up to this point we have considered only one type of relations, namely, comparative relations based on 'identity', diversity, and similarity. Another type of relationship which is of great significance in science is the causal directly connected; they are connected within and by the whole.

Dealing with relations and dealing with systems involve two different logical manipulations to which two psychologically quite different processes may correspond. In the recent past there has been much rather inconclusive discussion concerning the possibility of two different processes of knowing: explanation and understanding. I am referring here to discussion of the problem, *erklärende und verstehende Psychologie*. The difference between the two concepts, as they have been used in the aforementioned discussion, is probably that explanation refers to relational thinking, understanding to system thinking. Relational thinking aims at the establishment of the direct connexion between two objects. For instance, in the study of causation one has to find for member a (effect) a second member b (cause) with which it is necessarily connected. In causal research the task is to single out from a multiplicity of data pairs of facts between which there is a necessary connexion. In system thinking the task is not to find direct relations between members but to find the superordinate system in which they are connected or to define the positional value of members relative to the superordinate system.

The preceding discussion is far from being complete and adequate, but it is hoped that our main thesis at least, that relations and systems are two different logical genera with distinctive characteristics, has been given a relationship. Causal relationship can be expressed as follows: if a occurs, b follows. Causal relationships may be in many respects different from comparative relationships, but they are alike in that they represent direct connexions between one member and another. Even complex causal relationships can be analysed into two-membered relationships. Where such analysis is logically not possible, we deal with system-connectedness.

On the basis of the foregoing discussion the differences between relations and systems can be summarized in a preliminary fashion as follows: (1) Relationships involve two and only two members (relata). Complex relationships can always be analysed into pairs of relata. Systems may involve an unspecified number of components, not analysable in certain respects into pairs of relata. (2) The relata enter into a relationship by virtue of their immanent attributes, while the constituents enter into a system-connexion, not through their immanent attributes but through the positional value which they have in the system. Secondary relations which are based on positional values of the relata can be established also between members of a system. But the system itself cannot be described even in terms of such relationships. (3) For the existence of systems a dimensional domain is necessary. Systems are specific forms of the distribution of members in a dimensional domain. (4) In relationships the connexion between the relata is a direct one. The members of a system, on the contrary, do not need to be measures of plausibility. It is still an open question whether relations and systems are *absolutely* different logical genera, or whether the one may be a subtype of the other. The latter possibility cannot be excluded and there are certain arguments in favor of it. One thing, however, seems clear, namely, that systems cannot be deduced from relations, while the deduction of relations from systems still remains a possibility. If that is the case then the more general logical genus would be 'system', while 'relation' would be a reduced, simplified system which is adequate only for the logical presentation of very simple specialized constellations.

System and Gestalt

In the course of the past two decades it has been almost generally recognized by biologists and psychologists that the clarification of the problem of wholes is essential for progress in the study of the organism. The increasing awareness of the problem of wholes led to the discovery of certain general principles, best formulated perhaps by the Gestalt psychologists. It will be useful to examine briefly these formulations in the light of the previous discussion.

The most generally known thesis with regard to wholes is the following: 'The whole is more than the sum of its parts.' This is not a very felicitous formulation since – contrary to the concept of Gestalt psychologists – it may suggest that a summation of parts takes place and that, besides the summation, a new additional factor enters into the constitution of wholes. In Ehrenfels' *Gestaltqualität* such an additional factor actually has been suggested.

Wholes, however, cannot be compared to additive aggregations at all. Instead of stating that in the formation of wholes something more than a summation of parts takes place, it would be more correct to state that summation does not play any part whatsoever in the formation of wholes. In summations the parts function because of their inherent qualities. When, for instance linear distances are added to form a larger linear distance, the first distance, as such, is directly joined (*und-Verbindung*) to the second distance and this to the third, and so on. On the other hand, when a number of parts constitute a whole, the parts do not enter into such a connexion by means of their inherent qualities, but by means of their position in the system. The formation of wholes is therefore not additional to the aggregation of parts, but something of an entirely different order. *In aggregates it is significant that the parts are added; in a system it is significant that the parts are arranged.*

It should also be kept in mind that 'part' means something different when applied to aggregate from what it means when applied to wholes. When the single objects a, b, c, d, are bound together in an aggregate they participate in that aggregation as object a, object b, object c, etc., that is, as lines, distances, color spots, or whatever they may be. When, however, a whole is constituted by the utilization of objects a, b, c, d, the parts of the resulting whole are *not* object a, object b, object c, etc., but α, β, γ, δ, that is, *the positional values of the objects* a, b, c, d.

I would suggest that the principle 'the whole is more than the sum of its parts' be modified in the following way: aggregation and whole formation are processes of an entirely different order. And we may also formulate this statement more specifically: in an aggregation the parts are added, in wholes the parts are arranged in a system. The system cannot be derived from the parts; the system is an independent framework in which the parts are placed.

That the whole is, to a large extent, independent of the individual parts has been frequently pointed out. We may transpose a melody a few octaves higher or lower and it still remains essentially the same melody, although this transposition may be such that the two variations of the melody have no single individual tone in common. If we recall that this system is a kind of arrangement in which the parts do not participate by means of their inherent characteristics but by means of the positional values, the above-mentioned relative independence of the whole from the nature of the individual parts will be understandable.

The above statement needs, however, some qualification and restriction. The parts may need to have certain attributes which enable them to fill the positions which are required for the system. In a triangular geometrical arrangement the parts have to be *lines*, although their other properties (for instance, their absolute lengths) are irrelevant. Thus certain properties of the constituents are relevant, that is, they are necessary to permit the occupancy of a given position, while other properties are irrelevant. A similar distinction has been made also by J. von Uexküll, who differentiates between 'leading' and 'accompanying' properties (Uexküll, 1928). The greater the organization of the whole, the more the inherent properties of parts are utilized as co-determinants of positional values. The human organism, for example, is highly economical in this respect: it carries a minimal load of irrelevant properties of parts; most of the properties of parts are 'utilized', that is, are co-determinants of the positional value of the part.

'Wholes are never undifferentiated but always a *unitas multiplex*.' Let us place the emphasis first on the multiplicity. If one keeps in mind that the system is a way of arranging parts, the logical necessity of a multiplicity becomes evident since a single factor in itself cannot be arranged.

The term 'whole' is frequently used with a very confusing

double meaning. Sometimes the concrete organized *object*, other times the *organization* of the object is called a whole. The term is used in this latter connotation, for instance, when one states that a circle may be small or large, drawn in red or in green color, and still remain the same whole, the same Gestalt. I propose that the term 'whole' be reserved to designate the concrete *organized object*, while the *organization* itself, the way of arrangement of parts, should be called system.

The logical formulation of a given system states the construction principle or the *system principle* of the whole. Every system has one and only one construction principle. This is the meaning of the first term in the expression, *unitas multiplex*. The particular system principle may be perfectly or only approximately realized in a given whole. There are wholes in which all the significant positions of the system are occupied in perfect accordance with the system principle, but there are also wholes in which only a limited number of positions, sufficient to suggest the system principle, are occupied while other members are *out of position*. This is the difference which among Gestalt psychologists is somewhat vaguely referred to by the terms 'good' and 'bad' Gestalt. The various degrees of *Pregnanz* which a Gestalt may have express the degree of conformity of the positions of the parts with the system principle. There are also instances where in a whole a sufficient number of positions are occupied to indicate the system principle, while the other positions *are not filled*. These are 'open' Gestalts in contradistinction with 'closed' ones, wherein all significant positions are occupied.

Certain relations, as, for instance, comparative relations (a is larger than b), could be called static, while others could be called dynamic. The prototype of the latter is the causal relationship. In the same fashion one could distinguish between static and dynamic systems. We hope that we have demonstrated the existence of the static forms of systems. The question, however, as to the existence of dynamic systems still remains open. In static systems the whole imparts to its constituents a positional value which the given constituent does not have in itself, but only when it forms part of the given whole. With regard to dynamic wholes, one would expect that a given part *functions differently* depending on the whole to which it belongs. We would also expect that the whole has its own characteristic dynamics. Certain principles of holistic dynamics have already been formulated, as, for instance, the 'tendency to closure' and the 'tendency to *Pregnanz*'. Up to the present such dynamic principles have been satisfactorily demonstrated only within the psychological realm, which is generally characterized by great plasticity. That such principles hold true also in realms of greater rigidity is strongly suggested by certain facts, but it has not been satisfactorily demonstrated as yet. Such an assumption is a working hypothesis – and in the present discussion it is taken as such; a convincing demonstration remains a problem for future inquiry.

The possibility of the dynamic action of a system would probably be rejected *a priori* by many students. Although in the last analysis causality is just as inexplicable as a system action, still many students would feel more comfortable and would be willing to give credit for greater scientific validity to the formulation of the dynamics of a given happening in terms of causality than to its formulation in terms of system action. Causal thinking has been used in science for such a long time and, in certain fields, with such success that it is almost generally considered as *the* scientific thinking, although it may well be only a subvariety of it. Relational thinking is so firmly rooted a habit that the transition to system thinking is at least as difficult as the transition from a three-dimensional to a four-dimensional geometry.

This brief discussion of some holistic principles and concepts suggests the possibility of the logical formulation of some principles and concepts. Only a strict logical formulation can dispel the vagueness and obscurity which have been so common in the early holistic theories.

[. . .]

References

MEYER, A. (1935), *Ideen und Ideale der biologischen Erkenntnis*, Barth.
UEXKÜLL, J. VON (1928), *Theoretische Biologie*, Springer, 2nd edn.

2 J. Feibleman and J. W. Friend

The Structure and Function of Organization

J. Feibleman and J. W. Friend, 'The structure and function of organization', *Philosophical Review*, vol. 54 (1945), pp. 19–44.

In this essay we propose to set forth the structure and the function which are common to all empirical organizations, to serve as an instrument for the investigation of empirical organizations at every level. The aim of such a canon is to show not only the purposive functions of organizations, but primarily the composition which makes possible the fulfillment of such functions. We are therefore concerned first with static analysis. Then it will be seen how an understanding of the analysis of organization promotes a better comprehension of the interactions of organizations as wholes.

The study of organization, then, must be approached from two standpoints – that of statics and that of dynamics. Statics treats of organizations as independent of their environment and therefore as isolated from problems of interaction with other organizations. Dynamics treats of organizations as dependent to some extent upon their environment and therefore as interactive with other organizations. The division of the topic into statics and dynamics is of course not absolute. Actually, no organization is ever completely static or dynamic; all have structure and all suffer functional change. But statics and dynamics are logical and abstract divisions in terms of which the nature of organization can be determined. We cannot hope to understand the dynamics of organization without some prior understanding of statics.

I. Statics

The basis of organization

In treating of structure we first regard the organization itself as a whole. The whole obviously analyses into parts. These parts themselves have parts, which we shall term subparts. Thus there are two levels of analysis:

wholes (from which the analysis is made),
parts and subparts.

Wholes need no detailed description. Anything is a whole which operates in quasi-independence of its environment. A rock is a whole, and so is a mammal, an apple, a committee. Wholes are not a level of analysis but that from which analysis starts. Parts are the first level of analysis. All that need be said of them here is that they are the immediate basic factors into which wholes can be analysed, and that they are of a certain quantity. A rock analyses into such parts as crystals, an apple into such parts as skin, flesh, and seeds. Subparts are the second level of analysis. They are the immediate factors into which parts can be analysed. A crystal analyses into molecules, apple flesh into cells. It should be noted that subparts are parts of parts in the same sense in which parts are parts of wholes. Of course it is true that wholes themselves are parts of still larger wholes, and that subparts may be analysed into sub-subparts. For example, rocks are parts of mountains, apples of the tree-system, and molecules may analyse further into atoms, and cells into molecules. But it is our contention that for purposes of analysis of structure no more is required than the whole from which the analysis starts and two levels of analysis.

Elements of relations

There is another factor in the analysis of organization which we may temporarily describe as the ways in which parts exist in combination with other parts to form the structure of the whole. These 'ways of combination' are not basic in the same sense in which parts and subparts are basic. They are the kinds of relations between parts, the ways in which parts combine. Apple is not composed of flesh, skin, and seeds in any haphazard combination but in a definite set of relations between these. The elements of relations which exist between parts of an organization form a certain group of relations, listed as follows:

(a) transitivity	(-a) intransitivity
(b) connexity	(-b) non-connexity

(c) symmetry	(-c) asymmetry
(d) seriality (transitive, connected, asymmetrical)	(-d) aseriality
(e) correlation (one-many$_1$, one-one$_2$, many-one$_3$, many-many$_4$)	(-e) non-correlation
(f) addition	(-f) non-addition
(g) multiplication	(-g) non-multiplication
(h) commutation	(-h) non-commutation
(i) association	(-i) non-association
(j) distribution	(-j) non-distribution
(k) dependence	(-k) independence

There may be other relations between parts in any concrete organization – an indefinite number of others, in fact. But such relations do not bear on the question of organization *qua* organization. The relations listed above are the determinate ones for organization, those into which the relations of parts of an organization may be analysed without remainder. Before going further with our analysis we may define what is meant by these terms which, although commonplaces of logic, may appear strange in this context.

(a) Transitivity is such that if it relates two parts to a middle part, it relates the extreme parts to each other. (-a) Intransitivity is the absence of this relation. For example, the parts of an apple are transitive, for if the skin of an apple encloses the flesh, and the flesh encloses the seeds, then the skin encloses the seeds. An aggregation of three grains of sand is intransitive, since the extreme parts are not related to each other by a middle part.

(b) Connexity is the relation of two parts without the mediation of a third part. (-b) Non-connexity is the absence of this relation. For example, the skin and seeds of an apple are non-connected. The three grains of sand are connected.

(c) Symmetry is the relation in which the interchange of the parts does not involve any change in the relation. (-c) Asymmetry is the relation in which the interchange of the parts does involve a change in the relation. For example, the relation of the skin of an apple to the flesh is not the same as the relation of the flesh to the skin, and is therefore an asymmetrical relation. The three grains of sand are symmetrical since an interchange of parts does not involve any change in their relation.

(d) Seriality is the relation which is transitive, asymmetrical, and connected. (-d) Aseriality is the absence of any one of these relations. For example the enclosure of apple seeds by flesh and of flesh by skin is a serial relation (see a, b, and c above). The three grains of sand are aserial, since they lack asymmetry.

(e) Correlation (one–many$_1$, one–one$_2$, many–one$_3$, many–many$_4$) is the relation between two series such that for every part of one series there is a corresponding part in the other series and no part of either series is without a corresponding part in the other. All relations are examples of one type of correlation. For example, of two apples, the serial parts of one correlate with the serial parts of the second.

(f) Addition is the relation of the joining of parts so as to increase their number. For example, when the skin of an apple is added to flesh and seeds, there are three parts instead of two.

(g) Multiplication is the relation of the joining of parts so as to involve them with each other. For example, the relation of skin, flesh, and seeds in a whole apple is multiplicative rather than additive because these parts are involved with each other in ways other than mere aggregation.

(h) Commutation is the relation in which addition and/or multiplication is symmetrical. (-h) Non-commutation is the relation in which addition and/or multiplication is asymmetrical. For example, the relation of skin, flesh, and seeds is non-commutative, since the parts are multiplicative (see g above) only in serial order, i.e. they are asymmetrical (see d above). The three grains of sand are commutative, since their addition and/or multiplication is symmetrical.

(i) Association is the relation which is commutative and connected. (-i) Non-association is the absence of this relation. For example, the relation of skin, flesh, and seeds is non-associative, since it is non-commutative (see h) and association requires commutation. The three grains of sand are associative, since they are both commutative (see h) and connected (see b).

(j) Distribution is the relation which is commutative and intransitive. (-j) Non-distribution is the absence of this relation. For example, the relation of the skin, flesh, and seeds of an apple is non-distributive since skin, flesh, and seeds are non-commutative (see -h) and distribution requires commutation. The three grains of sand are distributive, since they are both commutative (see h) and intransitive (see a).

(k) Dependence is the relation in which the existence of one part is conditioned by some other part. (-k) Independence is the absence of this condition. For example, the limb of an animal is dependent upon the circulatory system, but not the circulatory system upon the limb. The three grains of sand are independent of each other. The heart and the circulatory system are interdependent.

Rules of organization

We now have the basis of organization, i.e. wholes, parts, and subparts, and we also have the elements of relations between parts. But these are not sufficient to define or determine any given organization. In addition to them we need certain rules in terms of which parts and their relations are constitutive of organizations. We shall list these rules and then discuss them. They are:

1. Structure is the sharing of subparts between parts.
2. Organization is the one controlling order of structure.
3. One more level is needed to constitute an organization than is contained in its parts and subparts.
4. In every organization there must be a serial relation.
5. All parts are shared parts.
6. Things in an organization which are related to parts of the organization are themselves parts of the organization.
7. Things in an organization which are related to related parts of the organization are themselves parts of the organization.
8. The number of parts and of their relations constitutes complexity.

1. The linkage of parts is accomplished by means of common subparts and not by mere juxtaposition or external linkage. The joining of two parts is effected by a subpart which they have in common, and this is the basis upon which all structures are constituted. For example, in carbon tetrachloride (whole) the carbon (part) and the chloride (part) share electrons (subparts).

2. It is not simply the fact of linkages but rather the principle according to which all linkages fall together into one controlling order, which makes an organization. For example, it is the arrangement of the four atoms of chlorine in a certain spatial

relation around one atom of carbon which constitutes the organization, carbon tetrachloride.

3. No number of subparts and parts constitutes an organization, which is essentially a property of the whole. The organization is one level above its analytic parts and subparts, and thus the whole involves another level. For example, carbon tetrachloride consists of something more than merely one carbon and four chlorine atoms.

4. The serial relation is the essential one in every organization. Other relations may and usually do exist in organizations but they are not necessary to the constitution of an organization. In the analysis of every whole there must be found a controlling relation which is asymmetrical, transitive, and connexive. For example, in carbon tetrachloride, the transitive relation is the relating of any two chlorine atoms to the carbon atom in such a way that they are related to each other. The asymmetrical relation is the fact that the carbon and chlorine atoms cannot be interchanged without destroying the carbon tetrachloride. The connexive relation is the relation of the carbon to the chlorine atoms without the intermediation of any third atom, i.e. of any external bond.

5. There is nothing in an organization except parts which have subparts in common. Any item which is not so shared is extraneous to the organization and not a part. For example, in carbon tetrachloride, all atoms of carbon and chloride are shared, and there is no unshared other atom.

6. Anything in an organization which is related to part of that organization is, by virtue of that relation, itself part of the organization, and not a foreign body. For example, any atom in carbon tetrachloride which is related to the carbon or chlorine atoms is itself a part of carbon tetrachloride.

7. Sometimes there are things in an organization which are not related to any single part of the organization, but which may be related to two or more parts. That is, the thing in the organization may be related to a complex of parts without being related to any one of the parts in the complex. In this case, the thing so related is part of the organization. For example, certain electrons which are required by carbon and chlorine in combination are themselves part of carbon tetrachloride, although they may not be required by either carbon or chlorine separately.

8. The number of parts and of their relations, i.e. subparts, constitutes the complexity of an organization. This rule and the yardstick of integrality, or kinds of organization [Statics, section (d)] form a pair of criteria. Complexity is seen to reduce to a mere matter of counting parts and subparts. For example, carbon tetrachloride has a complexity of five parts and thirty-two subparts.[1]

Kinds of organization

Having set forth the bare outline of the basis, elements, and rules of organization, we may now proceed to classify organizations as wholes into their various kinds. The kinds of organization constitute degrees of integrality, as follows:

(a) *Agglutinative.* The governing relation is aseriality, where parts have intransitivity, connexity and symmetry.
Relations are therefore -a, b, c, -d, e_4, f, -g, h, i, j.
(b) *Participative.* The governing relation is seriality. Participative organizations subdivide into three kinds, as follows:

(α) *Adjunctive.* The governing relation is symmetrical independence. The sharing of subparts is not necessary to either of the parts. Parts can survive their separation.
Relations are therefore a, b, -c, d, e_4, f, g, -h, -i, -j, -k.
(β) *Subjective.* The governing relation is asymmetrical dependence. The sharing of parts is necessary to one of the parts but not to both.
Relations are therefore a, b, -c, d, e_1 or e_3, -f, g, -h, -i, -j, k.
(γ) *Complemental.* The governing relation is symmetrical dependence. The sharing of parts is necessary to both of the parts. Neither part can survive separation.
Relations are therefore a, b, -c, d, e_3, -f, g, -h, -i, -j, k^2.

It should be understood that by 'Kinds of organization' is meant their logical and structural differences and not their division by qualitative or common-sense description. The division of organizations into various kinds is not arbitrary, although there are other classifications which could be made. The present one is made specifically according to the governing relations of the organizations. It needs to be pointed out that this classification is ideal in the sense that no actual organization ever completely

1. This count includes only the interactive, outermost shell of electrons. The inactive nuclei are not here considered.

answers to it.[2] However, since everything actual is an organization, everything actual has to fall under one or another, or under several, of the kinds of organization listed above.

The agglutinative organization is the loosest form of organization. In it the parts have only the spatial relation of contiguity. The more intimate relation which consists in the sharing of subparts by parts is absent here. Indeed there may be some question as to whether an agglutinative organization should be called an organization and not a mere aggregate of independent organizations. There is some justice in the contention that such an affair is not an organization, since it does not obey the Rules of Organization (see p. 34). It violates rules 1, 4, 5, and 6. We include the agglutinative as an organization only because of its illustrative value as a borderline case. An aggregate in some senses is an organization, but so tenuous a one that it can be easily dissipated and does not offer the kind of resistance which we may expect from an organization. Spatial relations are the least binding relations, yet they are not quite nothing, and to ignore them altogether is to overlook subtle relations of organization which presumably must have some effect.

One example of agglutinative 'organization' is a pile of sand whose grains may temporarily form a pattern. The same can be true of a crowd of people, a host of raindrops, gaseous molecules, a constellation of stars.

However, most of the organizations encountered are participative, where, besides spatial relations between parts, other relations enter to bind them closer together. In other words, in participative organizations parts are held together by the subparts which they share. The ways in which parts share subparts vary, and it is in terms of this difference that the participative classification lends itself to sub-classification. But all the sub-classifications have one essential property: they are all governed by the serial relation. The sub-classifications of participative organization are adjunctive, subjective, and complemental.

In adjunctive organizations parts share subparts, but this sharing is not so integral to either part that it cannot survive separation.

2. The term 'ideal' is employed here not as a synonym for mental or subjective but rather as a synonym for perfect. Cf. chemically pure iron, the ideal gas, etc.

Thus to a certain extent parts are independent of the organization of which they are parts, and this independence contributes to the flexibility of the whole organization. In the case of adjunctive organizations, the governing relation of symmetrical independence means that parts are on a parity with respect to their relations with other parts, and that neither is dependent upon the other in any necessary sense.

Adjunctive organizations may be exemplified as follows: the relation between moss and the tree on which it grows; the relation between fingernails or hair and the human organism; the relation of 'captured' planets to the bodies around which they revolve; the relation of a nation to the League of Nations.

In subjective[3] organizations parts share subparts, but this sharing though necessary to one of the parts is not necessary to the other. That is to say, one part can survive separation but the other cannot, and this contributes to the stubbornness of the whole organization. In the case of subjective organizations, the governing relation is asymmetrical dependence. This means that parts are not on a parity with respect to their relations with other parts, and that one is dependent upon the other in a necessary sense, but not vice versa.

Subjective organizations may be exemplified as follows: the relation between arm and body of the human organism; the relation between the rings and Saturn; the relation between honorary presidents and clubs.

In complemental organizations parts share subparts and this sharing is necessary to both the parts. That is to say, neither part can survive separation from the other, and this brings about the rigidity of the organization. In the case of complemental organizations, the governing relation is one of symmetrical dependence. This means that parts are on a parity with respect to their relations with other parts, and that each is dependent upon the other.

Complemental organizations may be exemplified as follows: the relation between the heart and the blood; the relation between the two parties of a two-party democracy; the relation between roots and bark of a tree.

3. The term 'subjective' is not employed here in its epistemological sense, as meaning the mental end-term of the knowledge relation, but rather as subject or subordinate.

II. Dynamics

In this section we shall approach the theory of organization from the dynamic standpoint. It will entail the consideration of organization no longer isolated from other organizations which go to constitute the environmental world of interaction. Here the abstract organization which we examined under Statics is further examined as it operates and is operated upon, i.e. functions with, things outside itself. The effect within upon parts (the strains) will be examined as it is occasioned from without by the interaction of wholes (the stresses). Or, conversely, the effect upon wholes (the stresses) will be examined as it is occasioned by the interaction of parts (the strains). Viewed either way, what is being considered is function.

Elements of interaction

The field of dynamics has to do primarily with certain conditions which prevail in the world of action and reaction. These may be described under the following terms:

1. Organization–Environment
2. Action–Reaction
3. Availability–Virtual indifference
4. Equilibrium–Disequilibrium
5. Saturation–Insufficiency–Superfluity
6. Flexibility–Rigidity
7. Stability–Instability

1. Organizations have been defined as wholes which operate in quasi-isolation from their environment. This is the static view. From the point of view of dynamics, wholes must be considered in their relation to environment. The relation of organization to environment is a reciprocal election. For example, fish dynamically have to be treated in relation with other fish and sea water, plankton, etc. Similarly, Boy Scout organizations cannot be considered without the environment of other social organizations, and the world at large.

2. Every actual organization is in constant change or motion. This change or motion is of two sorts. The environment changes the organization and the organization changes the environment.

Thus there is an action and a reaction effective in every instance of change. For example, the ocean affects the water-content of the atmosphere and the water-content of the atmosphere affects the ocean. The policy of the United States affects and is affected by the actions of the German government.

3. Availability is a characteristic of a limited part of the environment of an organization – that part which, determined by the nature of the organization, importantly affects and is affected by it. Virtual indifference is a characteristic of that less limited part of the environment of an organization – that part which, determined by the limited nature of the organization, hardly affects or is affected by it. There is no absolute line which can be drawn between availability and virtual indifference; they shade off into each other. Horses are importantly affected by their interaction with human beings, but hardly interactive with the gravitational field of the star Sirius. The veterans of foreign wars affect and are affected by the foreign policy of the United States, but are hardly affected by the religion of the Sikhs.

4. Equilibrium is the condition in which the influence exerted by the organization upon its available environment and the influence of the available environment upon the organization are in balance. Disequilibrium is the condition in which this balance does not exist. A minimum amount of equilibrium is a *sine qua non* of organization. For example, the tapeworm is so organized and in equilibrium with its environment in the intestinal tract that it resists the action of the pepsin and trypsin in which it is bathed. The family of European states is at present in a condition of violent disequilibrium.

5. Saturation is the condition of an organization in which all parts share and all subparts are shared. Insufficiency is the condition where all parts share and there are some unshared subparts. Superfluity is the condition where there are some unsharing parts and all subparts are shared. A saturated organization is one which hardly reacts with the available environment. It is more or less 'satisfied', and to that extent inert. There is an optimal pragmatic limit to the numbers of parts in an organization. An insufficient organization is one which is to be satisfied or saturated, and which is therefore elective. It can achieve this satisfaction only by interacting with the available environment, in

order to take on parts. A superfluous organization is oversaturated, and can achieve equilibrium only by interacting with the available environment in order to get rid of its superfluous parts. Therefore it may be said that insufficiency, saturation, and superfluity are comparatives in a scale of saturation: too little, enough, and too much. For example, argon is an 'inert' gas because it is saturated, i.e. all its electrons are shared and it has none to give up. Chlorine is an active gas because it is insufficient, i.e. it has not enough electrons. And lithium is an active element because it is superfluous, i.e. it has too many electrons.[4]

6. Flexibility is the capacity of an organization to suffer limited change without severe disorganization. Rigidity is the absence of this capacity. In a flexible organization the relations between the parts can be changed by the available environment by reaction without the destruction of the equilibrium. A flexible organization would of course have to be one in which the parts were not all complemental, although complemental relations may exist in flexible organizations. A rigid organization, on the other hand, must resist absolutely the action of the available environment, for if it cannot it falls into disequilibrium, i.e. it is destroyed as an organization. For example, a river is a flexible organization since its course and dimensions may be altered to a very great extent without damage to its essential structure. A vase is an example of a rigid organization, since it cannot suffer any deformation without destruction. A democracy is flexible in comparison with the rigidity of a theocracy. A democracy can change its policy radically and still remain a democracy, as for instance the United States, while a theocracy is held to a minimum amount of change and resists all efforts to alter it, as for instance Tibet.

7. Stability is the capacity of an organization to remain in equilibrium. Instability is the absence of this capacity. Of course stability is not unrelated to internal structure, as may be seen from the fact that stability may depend either upon the flexibility or the rigidity of the organization. That is, an organization may be stable either because it is flexible enough to meet the action–reaction demands of the available environment and thus preserve

4. For example, it is easier for lithium to give up one electron than it would be to take on seven in order to achieve its equilibrium which consists in the octet.

its equilibrium, or because it is so rigid that it fends off the action. A rock is an example of a relatively stable organization which is stable because it is rigid. A feather is an example of a relatively stable organization which is stable because it is flexible. The Roman Catholic Church is an example of a relatively stable social organization which is rigid. The Jewish social group is an example of a relatively stable social organization, which has been flexible enough to adapt itself readily to new conditions.

Rules of interaction

We now have the elements of interaction, but by themselves they are not adequate for dynamical analysis, and we need in addition rules to show how the elements function. These rules are:

1. Every organization elects some other organization or organizations.
2. In every action there is a sharing and an interchange.
3. All action is occasioned by the available environment.
4. Available environment is limited by interaction with organization.
5. All organizations strive toward equilibrium.
6. Saturated organizations remain unchanged.
7. Insufficient and superfluous organizations tend to change.
8. Flexibility is a condition of growth.
9. Rigidity is a condition of maintenance.

1. Dynamics pervades the organizational field. No organization is ever at rest or neutral toward all other organizations. It is impossible for any organization to be altogether without election. For example, a stone attracts and is attracted by the earth and every object on it. Chlorophyll strives to take energy from the sun's rays.

2. No action ever takes place without altering all those organizations which are involved in the action. This alteration consists in an interchange of parts and/or a sharing of subparts. The dynamical field is one in which an exchange is continually taking place. For example, two atoms may combine by sharing an electron, or two countries may exchange products.

3. Every strain in an organization is initiated by the stress. The largest change may take place within the organization, which may in turn react with the environment; but even in this case the initial disturbance must have occurred in the environment. Of course

this does not mean that the reaction of the organization is a mere mechanical response, equal and opposite to the stimulus. The response is inevitably dynamical. But although the organization is dynamical in its response, it is never absolutely initiative. For example, illness in the organism is occasioned by something from the environment, for example disease germs. Even so seemingly self-initiated a process as abstract thought is always occasioned by some outward stimulus however remote the stimulus may be from the train of thought it initiates.

4. We have said that available environment is that part of the environment which importantly affects or is affected by the organization. We can see now that by 'importantly affects' is meant interacts dynamically. Available environment is just as much affected by organization as organization is by available environment. For example, the cacti of the desert do not have as part of their available environment the moisture of the tropics because they are unable to interact with it; but they do affect the air of the desert. A book and a lamp on the same table do not share each other's available environment because they do not dynamically interact.[5]

5. Equilibrium is the ideal state to which all organizations aspire; its absolute attainment would mean permanence for the attaining organization. All interaction by organizations with their available environment conduces toward this ideal. For example, the biochemistry of the human organism maintains an equilibrium of organism and its waste products on the one hand and the available environment and its fresh supply on the other. A stone manages to maintain more or less equilibrium by virtue of its paucity of available environment, with which there is very little interchange.

6. An ideally saturated organization by definition could not interact with the environment since it has nothing to give to it and wants to take nothing from it. Therefore it must remain unchanged: an ideal condition which no organization ever completely attains. For example, argon is a relatively stable organization, as is also helium; in both chemical elements there are no free electrons to be shared. The Parsee social group is

5. Of course there is some interaction, i.e. that which takes place because of the gravitational field; but this is a case of virtual indifference.

relatively more saturated than the Christian Scientist group, because it has less to give and needs less.

7. Under- and over-saturated organizations react with the available environment in the effort to gain an equilibrium. Because they seek a condition where they will be safe from change, they are forced to cross a field of change. For example, labor unions are insufficient organizations which make strenuous efforts to gain further members. In chemistry a supersaturated solution of salt is an example of a superfluous organization, since the act of throwing down crystals is one of change.

8. It is apparent from the above that only insufficient organizations can grow, but there is another condition of growth which depends on the ability of the organization to suffer change and still preserve its essential integrity. The condition of flexibility depends upon the internal constitution of the organization, i.e. upon the fact that its parts are independent to some extent. This makes growth possible. For example, the young of any biological species are more flexible than the adults. Similarly, a church is more flexible in its formative stages than it is when its tenets have become fixed.

9. The more resistance an organization offers to the environmental changes imposed on it, the more permanent it is. The condition of rigidity depends upon the internal constitution of the organization, i.e. upon the fact that its parts are interdependent to some extent and closely knit. For example, bone is more closely knit than muscle and hence more rigid and permanent. The members of the family group are more interdependent than a political party; hence the family is more rigid.

Kinds of interaction

Having set forth the elements and the rules of interaction, we may now set forth the kinds of interaction in an attempt to show how the dynamic sequence of stimulus–response–effect operates in the relations between organization and environment.

Stimuli from the environment are either negligible, effective, or destructive. A negligible stimulus is one which is so slight that it provokes no organizational response. The negligible stimulus is a sub-threshold affair. Of course every stimulus obtains some response; but to say that it provokes no organizational response is to say that it produces no specific reaction of the organization as a whole. To brush past a penny scale provokes no response from the scale. A destructive stimulus is one which is so strong that it also produces no organizational response; in fact it destroys the organization. A high charge of gunpowder in a shotgun shell would destroy both shell and shotgun.

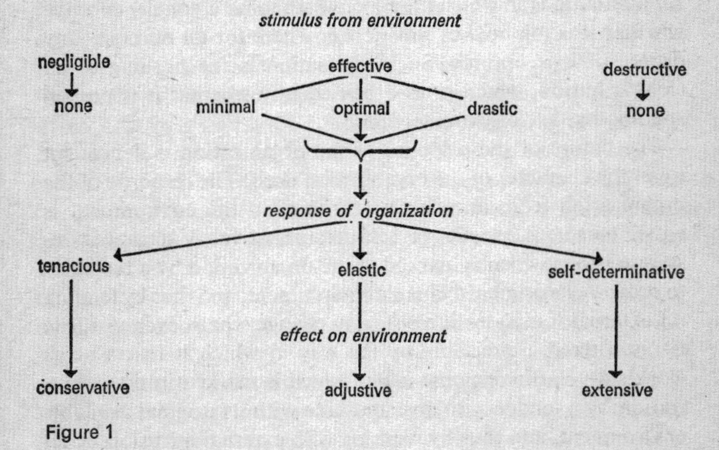

Figure 1

An effective stimulus is one which is strong enough to provoke an organizational response, but not too strong. It not only differs in degree from the negligible and destructive but it also differs in kind because it brings about a specific response which they cannot. To weigh on a penny scale, or to fire a shotgun with the properly charged shell is to provoke specific responses from the scale and the gun as whole organizations. Thus it is only the effective stimuli which concern us here.

Effective stimuli are divided into minimal, optimal, and drastic. A minimal stimulus is the least amount of stimulus which will provoke a response. A minimal stimulus reaches just over the threshold; any less stimulus than minimal is not responded to. A command barely heard provokes a response which is minimal since it may not be adequately responded to. An optimal stimulus is the stimulus which best provokes a full response. A command issued in a 'commanding voice' and fully audible will be adequately responded to, and therefore is an optimal stimulus. A drastic stimulus is the most amount of stimulus which will provoke a response without destruction. It is therefore barely less than a destructive stimulus. A command issued in such a stentorian voice that it antagonizes the one commanded might receive only a grudging and inadequate response. There are of course some situations in which the effective stimulus is equally effective whether it is minimal or drastic, i.e. where for all purposes any degree of it is optimal. Such a situation is the pushing of an electric button, which rings a bell equally whether it is pushed slightly, harder, or extremely hard.

The nature of the response of the organization is dependent upon the character of the organization itself. The response of the organization from an effective stimulus in the environment is either tenacious, elastic, or self-determinative. A tenacious response is one which is marked in the organization by a tendency to preserve its original available environment, and thus by fending off external intrusions it resists any change whatsoever. A turtle is not altered appreciably by the way in which it reacts to its world. An elastic response is one which is marked in the organization by a tendency to give and take with its original available environment, and thus by working with external intrusions it resists change. A tree that has the resiliency to give with the wind may preserve itself better than if it resisted. A self-determinative response is one which is marked in the organization by a tendency to change with its available environment and yet to remain itself by taking elements from the available environment and transforming them to suit itself. A horse is sensitive to new factors in his environment such as a stable mate, flies, a new master, and in some ways he makes responses which have effects upon these factors.

The effect on the environment of the response of the organization is either conservative or extensive. A tenacious response on the part of an organization produces a conservative effect on the environment. That is to say, it affects it as little as possible, and makes no alteration in its conditions. A turtle world of air and sea is not made very different by the way in which the turtle reacts to it. An elastic response on the part of an organization produces an adjustive effect on the environment. That is to say, it affects it somewhat and makes some alteration in its conditions. The tree which sways in the wind moderates the wind somewhat by giving with it. A self-determinative response on the part of an organization produces an extensive effect on the environment. That is to say, it affects it considerably and makes large alterations in its conditions. The horse which responds dynamically to stable mate, flies, and a new master exerts considerable effect upon these factors.

The conservative effect on the environment is that of resistance; whereas the adjustive and the extensive are those of cooperation. The first is blind; the second and third progressively perceptive. The area of available environment becomes wider with each of the three types of effect.

III. The Direction of Structure and Function

The terms, rules, and kinds of organization required for the analysis of structure and function, have been set forth. They have been set forth in tables and classifications, which attempt to show their interrelations. But the larger interrelations of structure and function have only been casually introduced. In this section we hope to show these interrelations by setting forth how function serves as the aim or end of structure.

This essay has proved to be a very abstract one. It does not study empirical fields of organization but rather that abstraction of organization which the empirical fields have in common. All conditions examined are ideal conditions, not in the sense that they consist of thoughts, but in the sense that they consist of abstractions existing as perfect logical possibilities. In order to make this more understandable, we shall abstract further from these abstractions to that ideal of organization toward which all actual organizations tend.

It is hoped that the essay as a whole can then serve as a canon for the examination of actual organizations in the various empirical scientific fields, and also for an examination of the work which has been done in the specific sciences.

Relation of statics to dynamics

In order to relate statics to dynamics, it will be necessary to show the relations between rules and kinds of organization on the one hand and rules and kinds of interaction on the other. It might be expected that we should also be able to relate the elements of relations with the elements of interaction. But of course it does not follow that because items combine in a related way that items themselves resemble each other. The parts of a structure do not necessarily resemble the different ways in which the structure functions. The atoms of the water molecule in no wise resemble the uses to which water can be put. Let us then pass on to the comparison of (α) rules and (β) kinds.

(α) *Static rule 1:* Structure is the sharing of subparts between parts. *Dynamic rule 2:* In every action there is a sharing and an interchange.

The sharing which constitutes the structure is from the point of view of the whole represented by the sharing which constitutes the function. The spongy structure of the lungs of the animal which share blood with other organs also enable the lungs to share and interchange gases with the atmospheric environment.

Static rule 2: Organization is the one controlling order of structure. *Dynamic rule 1:* Every organization elects some other organization or organizations.

Just as the organization is defined by one pervasive order which holds together the parts as one structure, so the whole in its election for other wholes functions as an order which holds together the parts. Thus action or function as a whole constitutes the controlling order of structures. A bottle is defined in its relation to its parts as a structure by the way in which it acts as a container of liquid for purposes beyond it. Thus the employment of bottles as storage containers for liquid dominates the structure of the bottle itself just as much as do the particles of glass which go to make it up.

Static rule 3: One more level is needed to constitute an organization than is contained in its parts and subparts. *Dynamic rule 4:*

Available environment is limited by interaction with organization.

Subparts, parts, and wholes are the three levels necessary to constitute an organization. But a whole is delimited by the available environment, through its capability of interaction with it. The relation between parts and available environment defines the whole, or, whole is defined by function. Just as available environment is limited by interaction with organization, part is limited by interaction with whole, and subpart by interaction with part. A tide limits its water-content and is limited by the interaction of the water with the mass of the moon.

Static rule 4: In every organization there must be a serial relation. *Dynamic rule 5:* All organizations strive toward equilibrium.

In every serial relation there is an element of asymmetry which is reduced to order by transitivity and connexity. Since no organization is ever in complete equilibrium, the striving toward equilibrium is the equivalent of the reduction of asymmetry to order in seriality. In the human organism there are innumerable asymmetrical relations; for instance the relations of the left foot to the right, the heart to the kidneys, the blood to the blood vessels, which are resolved into more or less order. Nevertheless, they cannot be resolved entirely, and the effort to resolve them is the struggle for equilibrium within an environment.

Static rule 5: All parts are shared parts. *Dynamic rule 6:* Saturated organizations remain unchanged.

In the perfect organization there are no parts which are not shared. This means that all subparts must be taken up or connected by other parts and that there can be no unshared parts which require anything from the available environment. But this is the condition of saturation. Thus, no unshared parts means – no change. In argon there are no unshared electrons and thus the gas is inert.

Static rule 6: Things in an organization which are related to parts of the organization are themselves parts of the organization. *Dynamic rule 3:* All action is occasioned by the available environment.

Things in the available environment which can be related to parts in an organization tend to be taken in by the organization to become themselves parts of the organization. But by being available to the organization, these things stimulate the organization to take them in. The sight and the smell of food stimulates the animal to incorporate the food into its own organization.

Static rule 7: Things in an organization which are related to related parts of the organization are themselves parts of the organization. *Dynamic rules 7, 8:* Insufficient and superfluous organizations tend to change. Flexibility is a condition of growth.

Obviously, static rule 7 is only an extension of static rule 6, whereas dynamic rules 7 and 8 appear to differ from dynamic rule 3, which was correlated with static rule 6, above. But if dynamic rules 7 and 8 be restated in the language of dynamic rule 3, this seeming disparity is dissipated. Action is occasioned by the available environment only in the case of insufficient and superfluous organizations. To act in response to the available environment, an organization must be able to grow. Such an organization which is capable of interaction is flexible.

The requirement of things in the available environment by parts of an organization is not limited to single parts but pervades the organization in the sense that related parts require what single parts do not, so that the complexity of parts increases the requirements. Thus the ability to form relations is a condition of growth. Those organizations in which there are some parts incompletely related are the organizations which are most likely to require things from the available environment and thus to grow. The blood is a flexible affair. It must get rid of superfluous carbon dioxide, and it is insufficient in oxygen. Thus it can grow only by interacting with the available environment of the atmosphere. Neither the histone, *globin*, nor the iron compound, *haem*, require oxygen in the same proportions in which haemoglobin requires oxygen.

Static rule 8: The number of parts and of their relations constitutes complexity. *Dynamic rule 9:* Rigidity is a condition of maintenance.

The more complex an organization, the more difficult is its maintenance. Complexity would appear to be in inverse relation to rigidity. We have already said that flexibility is a condition of growth (dynamic rule 8). Now, if we regard the complexity as a result of growth or evolution from less complex organizations, it will be seen why rigidity works against complexity. The amoeba is a more rigid organism than a giraffe. By the same token, it has maintained itself longer and it is less complex.

(β) *Kinds.* Kinds of organization and kinds of interaction are not to be simply correlated, since each one has to do with a different dimension of organization, and whole organizations must include both dimensions.

Taken from the point of view of complexity, the adjunctive kind of organization, where the sharing of subparts is not necessary to either of the parts, is less complex than the subjective, where the sharing of parts is necessary to one of the parts but not both, and these two types are still less complex than the complemental, where the sharing of the parts is necessary to both of the parts. Similarly, tenacious responses are less complex than elastic responses, and both are less complex than self-determinative responses. Taken from the point of view of integrality, the reverse proves to be the case – the adjunctive kind of organization is the least integral, the subjective more integral than the adjunctive but less integral than the complemental. In the case of the responses of the organization, however, it is the tenacious which is the most integral, the elastic which is less integral, and the self-determinative which is still less integral. A stone is less complex than a tree, and a tree less complex than a man. But a stone is more integral than a tree, and a tree more integral than a man.

It will now be clear how the criteria of organization, complexity and integrality, relate the kinds of organization with the kinds of interaction.

The perfect organization

We have related statics and dynamics, and in order to further clarify these relations, and further integrate them, it may be instructive to sketch the conditions with which a perfect organization would have to comply. Of course no such perfect organization has ever existed nor presumably can exist. But the theoretical plan of such an organization serves a very useful purpose in studying the direction of actual organizations. We can see the difficulties in the path of organization only when we know what the ideal of organization is.

Organizations do not contain within themselves and independent of their environment the urge to increase and to grow more perfect. What they do contain within themselves might be called almost the opposite of an urge. This is the tendency to retain an equilibrium forever. It is a conservative rather than a kinetic principle. But the environment in which every organization is involved will never let the organization altogether alone. It constantly stimulates and excites it by impinging upon it in more or less degree. This available environment forces the organization to make changes or perish. It is the interaction between the conservative tendency and external disturbances which causes various changes in the organization – changes taking several directions. These directions may be termed the three dodges of the organization to escape destruction. There is of course a fourth possible result which consists in the utter destruction of the organization, but that requires no further description.

The three dodges of the organization to survive are by tenacity, by elasticity, and by self-determination. Of course, the choice of these three dodges depends upon the structure of the organization itself. If the organization is too highly integrated for what it integrates, then it would tend to choose the path of tenacity. If, on the other hand, it is so complex that it has difficulties integrating what it contains, then it would tend to choose the path of self-determination. By choosing what it needs from the environment and so increasing its complexity in order to obtain better integration, it chooses the path of self-determination, a never-ending process. Between these two extremes lies elasticity, the compromise between integrality and complexity. Elasticity is not so integral that it cannot increase its complexity and not so complex that it cannot hold onto its integrality.

In order to illustrate the three dodges of the organization in response to the stimulus of the environment, we may take three examples which exist in the same general environment. The three examples are: a rock, some kelp, and a porpoise, and the environment is the ocean. The rock neither gives nor interacts with the ocean any more than can be helped. Water wears it away and it does nothing to build itself back, and thus it relies merely upon its integrality to continue its existence. The rock therefore has taken the tenacious dodge. Although the kelp to some extent resists outside influences as does the rock, and although it is to some extent self-determinative, it relies mainly on its elasticity to preserve its integrality. It allows itself to give and to yield to the conformations imposed upon it by the tides, the currents, etc., but it yet manages to interact with the ocean in such a manner as to preserve itself. It has taken the dodge of elasticity. Finally there is the porpoise which is able to control to a great extent its interactions with the medium in which it is bathed. It can go after food or not, it can swim against tide and current, it has an extra-survival margin, i.e. its time and energy are not altogether exhausted in maintaining its existence. It has taken the dodge of self-determination.

The dodges of tenacity and elasticity may seem to be at least as good as that of self-determination, in so far as mere continuance of existence is concerned. Indeed they may be better. But in terms of the perfect organization they have settled the problem of integrality at too low a level; they have compromised too soon. They have therefore no chance of ever becoming perfect. They can never grow in complexity, and so their integrality has too little to integrate. Self-determination on the other hand is a dodge which seems indefinitely capable of improvement, and this dodge alone gives the opportunity for working toward the ideal.

However, the self-determinative organization, by its very ability to grow, has to sacrifice a large amount of its integrality. Whereas a perfect organization would have complete integrality as well as complete complexity. Since complete integrality and complexity would mean that the perfect organization would contain everything and organize that everything completely, it is of course an ideal and not an actual thing; that is, its existence is infinitely remote.

Let us examine, then, the meaning of perfection in connection with organization. What would be the conditions for a perfect organization? Of course the first conditions stated above, that of perfect integrality and complexity, would mean an infinitely complex organization having no environment. Logically, this rules out dynamics, which is based upon the interaction of the organization with the environment. We shall have to choose the relations of the perfect organization from among those categories which have already been set up under statics.

The perfect organization is a whole containing an infinite number of parts and subparts and sub-subparts, etc., all completely organized. Since the final whole cannot by definition be related to anything else, its elements of relations are all negative; it is intransitive, non-connected, symmetrical, non-serial, non-correlative, non-commutative, non-associative, non-distributive, non-additive, non-multiplicative, and independent. It contains all positive relations; all transitivity, all connexity, all asymmetry, all seriality, all correlation, all commutation, all association, all distribution, all addition, all multiplication, and all dependence. In this connection it may be noted that each actual organization must contain some (but not all) of the elements of relations of which the perfect organizations contains all.

The perfect organization is complemental, since all other kinds of organization are incompletely integrated and therefore leave something to be desired.

We have stated that by definition the perfect organization has no environment and cannot interact. This renders the dynamic categories inapplicable, yet it may be of some interest to see how the dynamic categories are nullified when applied to the perfect organization. Referring to our elements of interaction (p.39) in 1, organization is everything, environment nothing; in 2, there is neither action nor reaction; in 3, there is no availability and hence virtual indifference becomes absolute indifference; in 4, the equilibrium is perfect; in 5, the saturation is complete, thus ruling out insufficiency and superfluity; in 6, it is as true to say of the perfect organization that it is flexible as to say that it is rigid; in 7, stability is absolute, since integrality is perfect.

The picture which we have drawn of a perfect organization is of course intended to be merely a suggestive one. Actually, there is no such organization known, and, due to the inherent limitations of the nature of knowledge, no such organization is likely to become known in the future. The purpose in setting up the conditions for such an organization is only to indicate the direction which actual organizations must take when they respond positively to stimuli from the environment. The diagram of the perfect organization exhibits the difficulties which every actual organization encounters in reconciling the demand for integrality with the corresponding stimulation toward complexity. So far as can be imagined, such a resolution will always be a compromise, more or less satisfactory and never altogether permanent.

We hope to have established a canon for the structure and function of organization abstracted from any and every empirical field and science. It will remain for further research to apply this canon to the empirical fields which the sciences study, and to the abstract sciences themselves.

Part Two **Properties of Open Systems**

Systems theory began with the recognition that certain principles held good for material systems despite the similarity or difference of their component parts. Reading 3 by Koehler well represents the early attempts to state the manner in which system properties regulate the behaviour of the components and hence the behaviour of the system. Many of these statements, like Angyal's in Reading 1, seem intuitively valid and important but it took von Bertalanffy's rigorous distinction between open and closed systems to mobilize widespread scientific interest.

From von Bertalanffy's classical paper (Reading 4) there arose the movement of thought that has sustained the *General Systems Yearbook* for more than a decade. Kremyanskiy wisely, and in our view correctly, questions the grounds and the aspirations of general systems theorizing. Granted the importance of system properties it does not follow that maximum scientific pay-off comes from seeking to identify the most general system properties. To pursue this goal is to run the risk of masking the environmental characteristics and the nature of the component parts that enter into the adaptive success or failure of concrete systems or classes of concrete systems. As a corrective to this editorial bias the reader is strongly advised to consult Miller (1965). This is the most exhaustive statement to date of the results that follow from a general systems approach. Even if we are right in our general appreciation of this approach there can be no doubt but that Miller has proven the value of general systems 'theory' in generating new hunches and hypotheses.

The efforts to formulate a scientifically adequate theory of living systems, both individual and group, came to fruition in Gerd Sommerhoff (1950).

We present here a later paper (Reading 6) which clears up some of the conceptual ambiguities brought out by the critical studies of the group of logicians centred around Ernest Nagel at the University of Columbia (Beckner, 1959). Nagel's own paper in which he demonstrates the central role of Sommerhoff's theory for the social sciences is presented in Part Four as Reading 15.

If it seems a little strange to the reader that an adequate solution to the theoretical problems of the 'properties of open systems' existed almost before the problem was posed (and thence generally ignored as if it did not exist), then he might also consider the fact that the logical structure for Sommerhoff's solution had been publicly formulated several years before by the American philosopher, E. A. Singer (for an equally appreciative audience).

References

BECKNER, M. (1959), *The Biological Way of Thought*, Columbia University Press.

MILLER, J. G. (1965), 'Living systems: cross level hypothesis', *Behavioral Science*, vol. 10, pp. 380–411.

SINGER, E. A. (1946), 'Mechanism, vitalism, naturalism', *Philosophy of Science*, vol. 13, pp.81–99.

SOMMERHOFF, G. (1950), *Analytical Biology*, Oxford University Press.

3 W. Koehler

Closed and Open Systems

Excerpt from chapter 8 of W. Koehler, *The Place of Values in the World of Fact*, Liveright, 1938, pp. 314–28.

[. . .] As it became more and more apparent that the machine principle is not capable of giving us a satisfactory explanation of organic regulation, an interpretation in more functional or dynamic terms began to attract some theorists. At first it seems indeed a plausible assumption that in the organism fitting regulation toward a standard status occurs for the same reasons that make physical systems attain or re-establish an equilibrium. Unfortunately, however, the concept of 'equilibrium' is in this connexion often used in just as vague a meaning as had previously been the case with the concept 'machine'. It appeared therefore advisable to analyse physical regulation before a comparison was undertaken between the normal state of an organism and an equilibrium in physics.

On the face of it, these standard states seem to resemble each other in a most promising manner. There is besides a special point which gives an equilibrium theory of organic regulation a particularly inviting appearance. Physical systems, we have found, tend to transform themselves in the direction of an equilibrium for two reasons: either because their processes follow the second law of thermodynamics, or because the law of dynamic direction applies to them. Now, even the most superficial consideration of the organism must convince everybody that its normal state cannot be a mere thermodynamic equilibrium. If, therefore, an equilibrium theory of organic regulation is to be at all proposed, this can be done only with the premise that both the law of dynamic direction and the second law apply to the organism; in other words, that the organism regulates toward a balance of directed vectors no less than it does toward 'a most probable situation'. We have seen, however, that the law of dynamic direction does not determine what actually happens in a system, unless there is sufficient friction by which inert macroscopic velocities are eliminated. Is this condition fulfilled in an organism? Without any doubt it is. In the movements of our limbs and in circulation inert velocities may perhaps play a modest role. In the tissue, however, friction is as great as it is in the interior of any solution. Consequently there are no such velocities in the tissue. What happens here must, from the point of view of physics, follow either from the second law or from the law of dynamic direction. In the former case we could say with the physicists that changes will occur in the direction of 'higher probabilities'; in the second case, that displacements will be proportional to, and in the direction of, the vectors which happen to obtain at each point.

The fact that the organism contains many 'devices', i.e. relatively permanent conditions of function, implies no obstacle for an equilibrium theory of organic regulation. In a physical system there may be many constraints; and yet, within the limitations which are thus imposed upon its operations, the tendency toward an equilibrium will determine what actually happens. We have only to realize that the equilibrium in question will itself respect those limiting constraints. In this sense the principal idea of all machine theories is entirely compatible with the more dynamic or functional notions to which an equilibrium theory refers.

On the other hand, it is the existence of relatively rigid anatomical conditions which restricts the range of possible organic regulations. Such devices as the organism possesses are undoubtedly apt to give the tendency toward standard states a general direction which is particularly fitting under more or less normal conditions. At the same time they exclude, precisely among the higher vertebrates, some regulations which might otherwise occur even in quite *un*usual situations. Such anatomical facts are, as it were, 'not made for these conditions'; and since they do not yield to the stress of altered function they prevent, under these circumstances, the actual occurrence of complete regulation.

So far, then, an equilibrium theory of regulation may seem to be wholly compatible with what we know about the organism. Any interpretation of organic fitness that does not take account of our two functional principles appears to me indeed as fundamentally unsound. On the other hand, it is equally true that neither the second law nor the law of dynamic direction can be applied to the organism in that simplest formulation in which they refer to equilibria. I have been at some pains to make the meaning of these principles more explicit than is often done, because I wish it to be perfectly clear that, unless a much broader view be taken, an equilibrium theory of organic regulation would be entirely misleading. To express the main argument against such a theory quite briefly: neither is the standard state of an organism a state of equilibrium in the common sense of the word, nor do organic processes in their totality generally tend to approach such an equilibrium.

In the introductory chapter of his book Cannon (1932) remarks that 'the constant conditions which are maintained in the body might be termed *equilibria*'. The author does not say what relation he assumes to obtain between this functional principle and his own view according to which regulation seems always to be due to regulating devices. At any rate, he prefers to give the name *homeostasis* to the fact that certain 'steady states' are so obstinately preserved or re-established in the organism. Equilibria, he adds, are found in simple closed systems, 'where known forces are balanced'. Again, he says, the word homeostasis 'does not imply something set and immobile, a stagnation'. There is something in these last words which many biologists may appreciate when attempts are made to explain organic regulation by an 'equilibrium theory'.

Convincing objections may, in fact, be raised against any such attempts. First, as Cannon says, no organism is detached from the rest of the world to an extent that would make our principles directly applicable to living systems. These systems are not closed. They absorb and they emit energy. At times they absorb much more than they emit. From the point of view of physics it is, therefore, simply impossible to state it as a rule that transformations in organisms occur in the direction of equilibria.

The same follows from the fact that in a healthy normal condition many vertebrates are by no means in equilibrium with regard to their immediate environment. Mammals, for instance, *stand* when at rest; for the most part they lie down only when slightly or seriously fatigued. Many fishes assume when at rest a position in which their heavier parts are turned away from the direction of gravitation. And yet in a state of physical equilibrium the center of gravity both of mammals and of fishes should be lowered as far as possible. Since no outer physical forces keep the mammals standing and the fishes swimming against the pull of gravitation, i.e. in an unstable position, such organisms must, when fresh, healthy, and therefore in their standard state, contain sets of vectors and processes which *prevent* the attainment of an equilibrium. These factors represent a certain amount of potential energy. But no physical system that is as such in a state of equilibrium can at the same time preserve an energy reserve by which it avoids reaching an equilibrium with regard to the environment. In the present example those factors seem even to keep the organism in a state *that departs from an equilibrium as much as possible.*

No conclusion other than this can be drawn from what happens during the development of individual organisms. During youth the standard state, for instance of man, varies slowly with time. There is always regulation toward a state that may be called temporarily normal. But from month to month, and from year to year, this state shifts gradually; and it is obvious that regulation changes its 'goal' correspondingly. Thus, if we know in what direction the standard state of the full-grown healthy adult differs from that of the healthy child, we shall also be enabled to define in what direction, toward what kind of state, regulation occurs when the individual is fully developed. We see the answer at once. In a state of equilibrium, as defined in this chapter, a system contains the smallest amount of potential energy which is compatible with given conditions. Potential energy, however, is the capacity of a system for macroscopic activities. Nobody, I am sure, will contend that a man of thirty can do less macroscopic work than can the new-born child. Just the contrary is true. From the point of view of physics the adult contains tremendous stores of potential energy when compared with the child. It follows that, in the healthy individual, development toward adult life is associated with an *increase* of such energy, that accordingly during this period regulation occurs in the direction of ever higher levels, and that, when development is at its peak, it is a *maximum*, not a minimum, of available energy which regulation tends to preserve.

These arguments must, I believe, convince everybody that an 'equilibrium theory' of organic homeostasis is not compatible with elementary biological facts. What is the theorist to do in this situation? In Professor A. V. Hill's words (1931, p. 60): '*If there be no equilibrium, how far dare we apply rules and formulae derived from the idea of equilibrium?*' In several statements the same author hints at a possible answer. All physiologists, he says, 'must have exercised their minds as to the reason why a living cell, completely at rest, and doing nothing at all except maintain its continued existence, requires a continual supply of energy' (Hill, 1931, p. 4). For instance, 'apart from any motor activity at all, a human muscle cell . . . uses, to maintain itself alone, about 30 calories of energy per gramme per day' (Hill, 1931, p. 60). We are, he finds, thus forced to adopt 'the conception of a dynamic steady state maintained by a continual expenditure of energy' (Hill, 1931, p. 62).

I should like to add that there is nothing hazardous in this conception. We can easily give it a clear functional meaning if we consider one more physical example. Life has sometimes been compared to a flame (e.g. Roux, 1914, p. 17, p. 79). This is more than a poetical metaphor, since, from the point of view of function and energetics, life and a flame have actually much in common. The flame, say, of a candle is a steady state. The continued existence of this state involves a continual supply of potential energy which the flame receives as 'food' through the wick and as oxygen from the air.[1] When undisturbed, the flame remains the same in size and in shape. Thus one might be tempted to believe that its status is that of an equilibrium. But in the sense in which this word is commonly used the flame is certainly not in a state of equilibrium. In order to see this we need only apply the same test to which we just subjected the adult organism: what is the genesis of a flame? We light a candle with a match. On the wick there appears at first a tiny flame. This flame grows spontaneously until it attains a maximum size and at the same time a certain shape, which then remain unaltered. If during the initial phase we hold our hand near the flame we can easily feel that quickly *increasing* amounts of heat are emitted at this time. We

1. Chemical energy is often treated in such cases as a form of potential energy. In fact, in all respects which concern us here it *is* potential energy.

also see that during this period the flame throws more and more light on its environment. Any energy, however, which the flame emits at a given moment was just before this moment inner energy of the flame itself; again, a moment before, it was potential (chemical) energy that was ready to be transformed into heat and light. From our simple observation it follows, therefore, that during its 'youth' the flame attains ever higher degrees of potential energy and that in its final stationary state it contains a maximum of such energy. In this sense the steady state of the flame departs as widely from a condition of equilibrium as it possibly can.

The factors which determine the maximum energy of the flame may be indicated in a few words. As soon as the candle is lighted, 'food' which is contained in the wick and oxygen that is contained in the air are being spent by combustion. Thus gradients are set up both for the food and for the oxygen. The flame begins to grow, and these gradients increase correspondingly. Higher amounts are therefore supplied both of food and of oxygen. But there is a limit to this process. When a certain size and a certain maximum of combustion have been reached, any further growth of the flame would lead to a higher demand than is compatible with the possible speed of oxygen diffusion from the surrounding air and with that of the food-stream which passes through the wick.

We are now in a position to apply our theoretical concepts to the flame and then to the organism. The flame is not a closed system. It can, however, be considered as part of a larger system for which our general principles are valid. If this be done certain consequences will follow for the behaviour of the flame as such. The air of the environment and the substance of the candle, taken together, contain amounts of chemical energy which are to all practical purposes unlimited. If therefore the material of the candle and a sufficiently large volume of air are included, we obtain a 'system' which we may regard as closed; because during the lifetime of the flame the energy on which its steady state depends will be exclusively supplied by the candle and this volume of air. An untrained observer's attention may be completely concentrated on the flame as an outstanding visual fact. From a functional point of view, however, the life of the flame can be understood only in the context of that larger 'system'. This system follows our general laws. The changes which occur in it must as a whole have a direction which lowers the amount of potential energy contained in the system. So long as the candle is not lighted this chemical energy cannot be spent at all. On the other hand, once a sufficiently high temperature is created at the tip of the wick, energy begins to be spent by combustion. And the more the flame grows, the more energy will be expended per unit of time.

It comes then to this: our system consists of, first, a practically unlimited store of chemical energy which, however, cannot be directly spent; and, secondly, of a minor part, the flame, in which this energy *can* be spent up to a certain maximum rate. The 'system' as a whole will lose its potential energy the more quickly, the more of this energy streams into the only part of it in which it can be expended. This is the flame. For this reason a maximum of such energy migrates steadily into the flame; for the same reason the flame *contains* continually a maximum of potential energy. Only thus is it enabled to expend energy at a maximum rate.

The fact that the stationary state of the minor system involves a maximum of potential energy is entirely compatible with our general principles. These refer to closed systems; and in the closed system of the present example, taken as a whole, events have precisely the direction which is prescribed by those principles. On the other hand, it is obviously an essential observation that in the only 'working' part of this closed system the direction of events is just the opposite of that to which the principles refer. The activity of the flame as such, it is true, namely combustion, tends to lower the amount of potential energy in the flame. But this does not really happen because any spent energy is at once replaced by a corresponding new supply. The more energy the flame emits, the higher is the rate at which it is supplied. Thus the flame is continuously fed with the greatest possible amount of energy.

A general view of the organism shows us a situation which resembles strongly that of the flame. The organism is not a closed system; it is part of a larger functional context, the external section of which contains as its most important components oxygen and food, i.e. a store of chemical energy which may be regarded as practically unlimited. In one respect there is a difference between the flame and the organism: unlike the flame, the organism itself normally contains great reserves of food in the widest sense of the word; it is stored, for instance, in the liver. From these sources rather than from the outside other tissues receive their food supply directly.

The potential energy of oxygen and food is not spontaneously spent outside the organism; nor is the food reserve consumed where it is stored within the organism. All 'activities', however, of which the organism is capable do tend to lower the supply of chemical energy that is contained in the active tissues. This is in line with our principles. But it is also in line with these principles that under such circumstances the stores of food deliver new supplies, and that these tend to maintain or to re-establish the highest energetic level of the tissues. If we compare this situation with that obtaining in a flame, we shall expect the active organism to heighten its content of potential energy during youth, and to preserve this content when a maximum is reached. This is exactly the behavior of the organism to which I pointed when I showed that the standard state of the organism cannot be a state of equilibrium.

Suppose now that at a given time there exists in the organism only one state of the tissue which corresponds to a maximum of potential energy. If by considerable work or by any other influence an organ or some larger part is changed so that for the organism as a whole the maximum condition is no longer maintained, such processes will occur as will bring it back to higher levels of energy. And since there is only one standard state in which a further increase is not possible, the organism will from various initial states 'regulate' toward that maximum condition. In other words, so far as regulation is concerned, our previous discussion of standard states applies to a maximum condition, just as it did to an equilibrium. Of course, it applies quite generally. There will, for instance, be in this case the same influence of given anatomical constraints as was mentioned in the case of regulation toward an equilibrium. Again, regulation toward a highest standard state will only be possible within limits which are given by relatively unalterable anatomical facts.

Actual regulation, however, will now be characterized by one remarkable trait. On the highest possible level of potential energy a system is capable of doing more macroscopic work than it is on any other level. If, therefore, in the organism any change by which this level is lowered tends to be followed by processes which counteract that change, and thus re-establish the highest level, regulation will serve to keep the living system in its most powerful state and, in this sense, to protect it. This, it seems, is the condition in which the various tissues are maintained by a constant supply of energy, and which is so often spontaneously restored after disturbances.[2]

I realize that in the last paragraphs no more than a general outline has been given, which cannot become a theory until a great many biological facts have been considered from this point of view. Since this task does not belong to our present program I shall only add a few tentative remarks.

One might ask why with these premises an organism does not live forever. My answer would be that regulation has its limits for reasons which I mentioned before; and that, therefore, a great many influences are able to destroy life. It is quite as obvious that without a sufficient supply of food and of oxygen the level of life will soon sink. For the same reason organisms can die from exhaustion, but they seem also to deteriorate spontaneously when a critical age has been passed. It appears to me quite likely that practically all our activities tend slightly to alter the tissue in a way which does not at once disturb further function, but which cannot be fully compensated for by the metabolism and its regulatory tendency. If such changes accumulate for years they may gradually make the tissue or certain organs less fit to respire and to absorb food. What this would mean is fairly obvious.

A further question refers to the manner in which the organism obtains its food. So long as within the organism there are reserves

2. I have sometimes been asked why I refuse to call the standard state of the organism 'an equilibrium'. My reason is simply that this standard state is not an equilibrium in any sense which has as yet been defined by science – not even an *un*stable equilibrium. It is a *stationary process*; and we are just beginning to learn that there are two classes of stationary processes, one with which a minimum, and another with which a maximum of energy is associated. Nothing could be more unfortunate than an attempt to hide such new essential distinctions behind an outworn general term.

of carbohydrates, fats, and so forth, the situation seems simple enough. But these reserves are limited, and I want to make it quite clear that the way in which the organism replenishes its stores is something altogether different from the simple processes that feed a flame with chemical energy. Delivery of both food and oxygen is in the case of the flame entirely automatic and direct. The existence of the flame sets up gradients in both respects, and those materials move in the direction of the gradients. Not much oxygen penetrates in this simple manner into the interior of a bird or a mammal; nor does food migrate into our interior simply because we are spending energy there. In the case of food, for instance, what happens is more complicated: among our activities there is one group, that of finding and of eating food, in which we *expend* certain amounts of energy, just as we do in all the others. As a result of these activities, however, we *absorb* under favorable conditions much more potential energy than is spent in order to obtain it. Thus the organism stores new supplies which enable it to maintain its raised normal level and, besides, to be active in other directions. In this there is nothing that could unbalance the energy budget; there is no contradiction between such operations and our general principles. None the less it must be noted that this particular behavior cannot be *predicted* from such principles, and that, from a general functional standpoint, it is a remarkable trait of living systems, about as remarkable as their reproductive activities. An actual theory of the organism will, if I am not mistaken, meet its most fascinating problems here.

There is no apparent reason, however, why science should hesitate to deal with the problem of *regulation*. Nor is there any essential difficulty in the fact that organic regulation is generally directed toward an 'optimum' state. The physicists, it is true, have not given much attention to this functional possibility, although it follows from their general laws. Also, a thorough investigation of *heterogeneous* physical systems and of their regulatory behavior would contribute greatly to a further understanding of organic fitness. But even now we can predict from known principles that some such systems will show an impressive causal harmony by which they themselves keep in a 'healthy' condition, as though this were their goal.

It is perhaps too early for final statements in this field, particularly since we know so little about the way in which evolution has created the living world. As this world now is, however, the following seems to me a conservative description of the situation: no procedure of science reveals any actual participation of demands and values in the determination of organic events. At the same time, science can clearly demonstrate that in certain systems function will, for dynamic reasons, take a most 'fitting' course. We do not discover requiredness as such among the data of science. But a general trend of nature is sometimes found to yield the same results as might be expected if the events in question were actually happening in order to fulfil a demand.

References

CANNON, W. B. (1932), *The Wisdom of the Body*, Norton.

HILL, A. V. (1931), *Adventures in Biophysics*, University of Pennsylvania Press.

ROUX, W. (1914), *Die Selbstregulation*, Berlin.

4 L. von Bertalanffy

The Theory of Open Systems in Physics and Biology

L. von Bertalanffy, 'The theory of open systems in physics and biology', *Science*, vol. 111 (1950), pp. 23–9.

From the physical point of view, the characteristic state of the living organism is that of an open system. A system is closed if no material enters or leaves it; it is open if there is import and export and, therefore, change of the components. Living systems are open systems, maintaining themselves in exchange of materials with environment, and in continuous building up and breaking down of their components.

So far, physics and physical chemistry have been concerned almost exclusively with processes in closed reaction systems, leading to chemical equilibria. Chemical equilibria are found also in partial systems of the living organism – for example, the equilibrium between hemoglobin, oxyhemoglobin, and oxygen upon which oxygen transport by blood is based. The cell and the organism as a whole, however, do not comprise a closed system, and are never in true equilibrium, but in a steady state. We need, therefore, an extension and generalization of the principles of physics and physical chemistry, complementing the usual theory of reactions and equilibria in closed systems, and dealing with open systems, their steady states, and the principles governing them.

Though it is usual to speak of the organism as a 'dynamic equilibrium', only in recent years has theoretical and experimental investigation of open systems and steady states begun. The conception of the organism as an open system has been advanced by von Bertalanffy since 1932, and general kinetic principles and their biological implications have been developed (4, 6). In German literature, Dehlinger and Wertz (15), Bavink (1), Skrabal (31), and others have extended these conceptions. A basically similar treatment was given by Burton (12). The paper of Reiner and Spiegelman (28) seems to have been inspired by conversations of the present author with Reiner in 1937–8. Starting from problems of technological chemistry, the comparison of efficiency in batch and continuous reaction systems, Denbigh *et al.* (16) have also developed the kinetics of open reaction systems. The most important recent work is the thermodynamics of open systems by Prigogine (25, 26).

In physics, the theory of open systems leads to fundamentally new principles. It is indeed the more general theory, the restriction of kinetics and thermodynamics to closed systems concerning only a rather special case. In biology, it first of all accounts for many characteristics of living systems that have appeared to be in contradiction to the laws of physics, and have been considered hitherto as vitalistic features. Second, the consideration of organisms as open systems yields quantitative laws of important biological phenomena. So far, the consequences of the theory have been developed especially in respect to biological problems, but the concept will be important for other fields too, such as industrial chemistry and meteorology.

General Characteristics of Open Systems

Some peculiarities of open reaction systems are obvious. A closed system *must*, according to the second law of thermodynamics, eventually attain a time-independent equilibrium state, with maximum entropy and minimum free energy, where the ratio between its phases remains constant. An open system *may* attain (certain conditions presupposed) a time-independent state where the system remains constant as a whole and in its phases, though there is a continuous flow of the component materials. This is called a steady state.[1] Chemical equilibria are based upon reversible reactions. Steady states are irreversible as a whole, and individual reactions concerned may be irreversible as well. A closed system in equilibrium does not need energy for its preservation, nor can energy be obtained from it. To perform work, however, the system must be, not in equilibrium, but tending to attain it. And to go on this way, the system must maintain a

1. In German, the term *Fliessgleichgewicht* was introduced by von Bertalanffy.

steady state. Therefore, the character of an open system is the necessary condition for the continuous working capacity of the organism.

To define open systems, we may use a general transport equation. Let Q_i be a measure of the *i*-th element of the system, e.g. a concentration of energy in a system of simultaneous equations. Its variation may be expressed by:

$$\frac{\partial Q_i}{\partial t} = T_i + P_i \qquad \mathbf{1}$$

P_i is the rate of production or destruction of the element Q_i at a certain point of space; it will have the form of a reaction equation. T_i represents the velocity of transport of Q_i at that point of space; in the simplest case, the T_i will be expressed by Fick's diffusion equation. A system defined by the system of equation **1** may have three kinds of solutions. First, there may be unlimited increase of the Q's; second, a time-independent steady state may be reached; third, there may be periodic solutions. In the case that a steady state is reached, the time-independent equation:

$$T_i + P_i = 0 \qquad \mathbf{2}$$

must hold for a time $t \neq 0$. If both members are linear in the Q_i and independent of t, the solution is of the form:

$$Q_i = Q_{i1}(x, y, z) + Q_{i2}(x, y, z, t) \qquad \mathbf{3}$$

where Q_{i2} is a function of time decreasing to 0 for certain limiting conditions.

We may consider the following simple case of an open system (4, 6). Let there be a transport of material a_1 into the system which is proportional to the difference between its concentrations X outside and x_1 inside $(X - x_1)$. This imported material may form, in a monomolecular and reversible reaction, a compound a_2 of concentration x_2. On the other hand, the substance a_1 may be catabolized, in an irreversible reaction, into a_3. And a_3 may be removed from the system, proportional to its concentration x_3. Then we have the following system of reactions:

$$\begin{array}{c} K \quad k_1 \\ \rightarrow a_1 \rightleftarrows a_2 \\ k_3 \downarrow k_2 \\ a_3 \\ \kappa \downarrow \\ \text{outflow} \end{array}$$

and equations:

$$\frac{dx_1}{dt} = K(X - x_1) - k_1x_1 + k_2x_2 - k_3x_1 = x_1 (-K - k_1 - k_3) + k_2x_2 + KX$$

$$\frac{dx_2}{dt} = k_1x_1 - k_2x_2 \qquad 4$$

$$\frac{dx_3}{dt} = k_3x_1 - \kappa x_3$$

Solving these equations for the steady state by equating to 0 we obtain that:

$$x_1 : x_2 : x_3 = 1 : \frac{k_1}{k_2} : \frac{k_3}{\kappa} \qquad 5$$

Immediately we see some interesting consequences. First, the composition of the system in the steady state remains constant, though the ratio of the components is not based upon a chemical equilibrium of reversible reactions, but the reactions are going on and are, in part, irreversible. Second, the steady state ratio of the components depends only on the system constants, not on the environmental conditions as represented by the concentration X. Third, we find $x_1 = \dfrac{KX}{K + k_3}$. Let us assume that a disturbance from outside, i.e. biologically speaking, a 'stimulus', raises the rate of catabolism, amounting to an increase of constant k_3. Then x_1 must decrease. But since import is proportional to the difference of the concentrations $X - x_1$, influx must increase. The system therefore manifests forces which are directed against a disturbance of its steady state. In biological language, we may say that the system shows adaptation to a new situation. Physicochemically, open systems show a behavior which corresponds to the principle of Le Chatelier.

But these characteristics of steady states are exactly those of organic metabolism. In both cases, there is first maintenance of a constant ratio of the components in a continuous flow of materials. Second, the composition is independent of, and maintained constant in, a varying import of materials; this corresponds to the fact that even in varying nutrition and at different absolute sizes the composition of the organism remains constant. Third, after a disturbance, a stimulus, the system re-establishes its steady state. Thus, the basic characteristics of *self-regulation* are general properties of open systems.

The energy need for the maintenance of a steady state is, in the simplest case of a monomolecular reversible reaction $a \underset{k_2}{\overset{k_1}{\rightleftarrows}} b$:

$$\frac{dA}{dt} = \frac{\Gamma - K}{\Gamma} \kappa_1 x_a{}^* RT \ln \frac{\Gamma}{K} \qquad 6$$

where $x_a{}^*$ is the concentration in the steady state, κ_1 the reaction velocity in the steady state, $K = k_1/k_2$ the equilibrium constant, $\Gamma = \kappa_1/\kappa_2$ the steady state constant. A similar expression applies to systems of n reaction partners (6).

Though at present this expression can hardly be used quantitatively in respect to cell problems, some general considerations are not without interest. Even the resting cell, performing no perceptible work, needs a continuous energy supply, as demonstrated by the fact that a deprivation of oxygen stops life in the (aerobic) cell. This maintenance work of the living cell is, of course, partly physicochemical, as in maintenance of osmotic pressure, and of ion concentrations different from those in the environment (17). But the chemical side must also be taken into account. Apart from photosynthesis, which needs considerable energy that is yielded by sun radiation, the anabolism of cell materials consists mainly of processes of dehydrosynthesis, such as the building up of proteins from amino acids and of polysaccharides from monosaccharides. Considered from the usual static viewpoint, these processes need little energy. But the situation is different in that these processes are to be considered not merely as a reversal of hydrolytic splitting, but the high molecular compounds must be maintained in a concentration very far from equilibrium. In the catalytic reactions, sugars $\rightleftarrows$ polysaccharides, amino acids $\rightleftarrows$ proteins, the equilibrium is almost completely on the side of the splitting products, as shown simply by the ease of hydrolysis and the difficulty, even impossibility, of dehydrosynthesis *in vitro*. Therefore, the steady state constant Γ is very different from the equilibrium constant K (35). The energy need for synthesis is demonstrated by numerous facts showing the coupling of anabolic and oxidative processes (6, p. 218 ff.). On the other hand, the efficiency of the 'living machine' appears to be rather low. Since the organism works as a chemodynamic system, theoretically an efficiency of 100 per cent, i.e. complete transformation of free energy into effective work, would be possible in isothermic and reversible processes, the condition of isothermy being almost ideally realized in the living organism. But the efficiency of the organic system in performing effective work, except in photosynthesis, does not much surpass that of man-made thermic machines. It appears that we have to take into account, in the balance of cell work, not only effective work but also conservation energy, i.e. the energy needed for the maintenance of the steady state.

If these considerations are correct, it is to be expected that there is heat production in the transition from the steady state to equilibrium, i.e. in cell death. In contradiction to earlier work of Meyerhof, this is true according to Lepeschkin (20), whose investigations, however, need confirmation. Possibly a combined study of reaction heat in cell death and of oxygen consumption of the resting cell, and the application of equation **6** can lead to a deeper insight into the problem of maintenance work of the cell.

Equifinality

A profound difference between most inanimate and living systems can be expressed by the concept of *equifinality*. In most physical systems, the final state is determined by the initial conditions. Take, for instance, the motion in a planetary system where the positions at a time t are determined by those of a time t_0, or a chemical equilibrium where the final concentrations depend on the initial ones. If there is a change in either the initial conditions or the process, the final state is changed. Vital phenomena show a different behavior. Here, to a wide extent, the final state may be reached from different initial conditions and in different ways. Such behavior we call equifinal. It is well known that equifinality has been considered the main proof of vitalism. The same final result, namely, a typical organism, is achieved by a whole normal germ of the sea urchin, a half-germ, or two fused germs, or after translocations of cells. According to Driesch, this is inexplicable in physicochemical terms: this extraordinary performance is to be accomplished only by the action of a vitalistic factor, an entelechy, essentially different from physicochemical forces and governing the process by foresight of the goal to be reached. Therefore, it is a question of basic importance whether equifinality is a proof of vitalism. The answer is that it is not (4, 6).

Analysis shows that closed systems cannot behave equifinally. This is the reason why equifinality is, in general, not found in inanimate systems. But in open systems which are exchanging materials with the environment, in so far as they attain a steady state, the latter is independent of the initial conditions; it is equifinal. This is expressed by the fact that if a system of equations of the form **1** has a solution of the form **3** the initial conditions do not appear in the steady state. In an open reaction system, irrespective of the concentrations in the beginning or at any other time, the steady state values will always be the same, being determined only by the constants of reactions and of the inflow and outflow.

Equifinality can be formulated quantitatively in certain biological cases. Thus growth is equifinal: the same species-characteristic final size can be reached from different initial sizes (e.g. in litters of different numbers of individuals) or after a temporary suppression of growth (e.g. by a diet insufficient in quantity or in vitamins). According to quantitative theory (p. 73), growth can be considered the result of a counteraction of the anabolism and catabolism of building materials. In the most common type of growth, anabolism is a function of surface, catabolism of body mass. With increasing size, the surface–volume ratio is shifted in disfavor of surface. Therefore, eventually a balance between anabolism and catabolism is reached which is independent of the initial size and depends only on the species-specific ratio of the metabolic constants. It is, therefore, equifinal.

Equifinality is found also in certain inorganic systems which, necessarily, are open ones (6, 9, 10). Such systems show a paradoxical behavior, as if the system 'knew' of the final state which it has to attain in the future.

Thermodynamics of Open Systems

It has sometimes been maintained that the second law of thermodynamics does not hold in living nature. Remember the sorting demon, invented by Maxwell, and Auerbach's doctrine of ectropy, stating that life is an organization created to avert the menacing entropy-death of the universe. Ectropy does not exist. However, thermodynamics was concerned only with closed systems, and its extension to open systems leads to very unexpected results.

It has been emphasized by von Bertalanffy (4, 6) that, 'according to definition, the second law of thermodynamics applies only to closed systems, it does not define the steady state'. The extension and generalization of thermodynamical theory has been carried through by Prigogine (11, 25, 26, 30). As Prigogine states, 'classical thermodynamics is an admirable but *fragmentary* doctrine. This fragmentary character results from the fact that it is applicable only to states of equilibrium in closed systems. *It is necessary, therefore, to establish a broader theory, comprising states of non-equilibrium as well as those of equilibrium.*' Thermodynamics of irreversible processes and open systems leads to the solution of many problems where, as in electrochemistry, osmotic pressure, thermodiffusion, Thomson and Peltier effects, etc., classical theory proved to be insufficient. We are indicating only a few, in part revolutionary, consequences.

Entropy must increase in all irreversible processes. Therefore, the change in entropy in a closed system must always be positive. But in an open system, and especially in a living organism, not only is there entropy production owing to irreversible processes, but the organism feeds, to use an expression of Schrödinger's, from negative entropy, importing complex organic molecules, using their energy, and rendering back the simpler end products to the environment. Thus, living systems, maintaining themselves in a steady state by the importation of materials rich in free energy, can avoid the increase of entropy which cannot be averted in closed systems.

According to Prigogine, the total change of entropy in an open system can be written as follows:

$$dS = d_eS + d_iS, \qquad 7$$

d_eS denoting the change of entropy by import, d_iS the production of entropy due to irreversible processes in the system, like chemical reactions, diffusion, and heat transport. The term d_iS is always positive, according to the second law; d_eS, however, may be negative as well as positive. Therefore, the total change of entropy in an open system can be negative as well as positive. Though the second law is not violated, or more precisely, though it holds for the system plus its environment, it does not hold for the open system itself. According to Prigogine, we can therefore state that: (a) Steady states in open systems are not defined by maximum entropy, but by the approach of minimum entropy production. (b) Entropy may decrease in such systems. (c) The steady states with minimum entropy production are, in general, stable. Therefore, if one of the system variables is altered, the system manifests changes in the opposite direction. Thus, the principle of Le Chatelier holds, not only for closed, but also for open systems. (d) The consideration of irreversible phenomena leads to the conception of thermodynamic, as opposed to astronomical, time. The first is nonmetrical (i.e. not definable by length measurements), but arithmetical, since it is based upon the entropy of chemical reactions and, therefore, on the number of particles involved; it is statistical because based upon the second law; and it is local because it results from the processes at a certain point of space.

The significance of the second law can be expressed also in another way. It states that the general trend is directed toward states of maximum disorder and leveling down of differences, the higher forms of energy such as mechanical, chemical, and light energy being irreversibly degraded to heat, and heat gradients continually disappearing. Therefore, the universe approaches entropy death when all energy is converted into heat of low temperature, and the world process comes to an end. There may be exceptions to the second law in microphysical dimensions: in the interior of stars, at extremely high temperatures, higher atoms are built up from simpler ones, especially helium from hydrogen, these processes being the source of sun radiation. But on the macrophysical level, the general direction of events toward degradation seems to be the necessary consequence of the second law.

But here a striking contrast between inanimate and animate nature seems to exist. In organic development and evolution, a transition toward states of higher order and differentiation seems to occur. The tendency toward increasing complication has been indicated as a primary characteristic of the living, as opposed to inanimate, nature (2). This was called, by Woltereck, 'anamorphosis', and was often used as a vitalistic argument.

These problems acquire new aspects if we pass from closed systems, solely taken into account by classical thermodynamics, to open systems. Entropy may decrease in open systems. Therefore, such systems may spontaneously develop toward states of greater heterogeneity and complexity (11, 25, 26, 30). Probably it is just the thermodynamical characteristic of organisms as open systems that is at the basis of the apparent contrast of catamorphosis in inanimate, and anamorphosis in living, nature. This is obviously so for the transition toward higher complexity in development, which is possible only at the expense of energies won by oxidation and other energy-yielding processes. In regard to evolution, these considerations show that the supposed violation of physical laws (13) does not exist, or, more strictly speaking, that it disappears by the extension of physical theory.

As emphasized by Prigogine, 'the thermodynamics of irreversible phenomena is an indispensable complement to the great theories of macrophysics, giving the latter a unification hitherto lacking'. Not since de Vries' and Pfeffer's work on osmosis have basic developments in physical theory been instigated by biological considerations. Not only must biological theory be based upon physics; the new developments show that the biological point of view opens new pathways in physical theory as well.

Biological Applications

Generally speaking, the basic fundamental physiological phenomena can be considered to be consequences of the fact that organisms are quasi-stationary open systems. Metabolism is maintenance in a steady state. Irritability and autonomous activities are smaller waves of processes superimposed on the continuous flux of the system, irritability consisting in reversible disturbances, after which the system comes back to its steady state, and autonomous activities in periodic fluctuations. Finally, growth, development, senescence, and death represent the approach to, and slow changes of, the steady state. The theories of many physiological phenomena are, therefore, special cases of the general theory of open systems, and, conversely, this conception is an important step in the development of biology as an exact science. Only a few examples can be briefly mentioned.

Rashevsky's theoretical cell model (27), representing a metabolizing drop into which substances flow from outside and undergo chemical reactions, from which the reaction products flow out, is a simple case of an open system. From this highly simplified abstract model, consequences can be derived which correspond to essential characteristics of the living cell, such as growth and periodic division, the impossibility of a 'spontaneous generation', an order of magnitude similar to average cell size, and the possibility of nonspherical shapes.

Permeation of substances into the cell, leading to a composition of the cell sap different from that of the surrounding medium, to the selective accumulation of salts, and to volume increase, was studied by Osterhout and his co-workers (23, 24) in large plant cells and physical models. The conditions are again those of open systems, attaining a steady state. The mathematical treatment of this system, as given by Longworth (22), is interesting because it agrees with and confirms the inferences, drawn from quite different physiological considerations, in von Bertalanffy's theory of growth (6).

The conception of open systems has been applied, by Dehlinger and Wertz (15), to elementary self-multiplying biological units, i.e. to viruses, genes, and chromosomes. A more detailed model was indicated by von Bertalanffy (7). According to this model, self-duplication in viruses and genes results from the fact that they are metabolizing aperiodic crystals. If degradative processes are going on, then, according to the derivations of Rashevsky, repulsive forces must result which eventually can lead to division and self-multiplication. At least in respect to chromosomes, this conception appears to be well founded, since tracer investigations with radiophosphorus show that the nucleoproteids of the cell are continually worn out and regenerated.

Tracer studies of metabolism, in particular, helped in this country to popularize the conception of the organism as a steady state.

> The discovery and the description of the dynamic state of the living cells is the major contribution that the isotope technique has made to the field of biology and medicine ... The proteolytic and hydrolytic enzymes are continuously active in breaking down the proteins, the carbohydrates and lipids at a very rapid rate. The erosion of the cell structure is continuously being compensated by a group of synthetic reactions which rebuild the degradated structure. The adult cell maintains itself in a steady state not because of the absence of degradative reactions but because the synthetic and degradative reactions are proceeding at equal rates. The net result appears to be an absence of reactions in the normal state; the approach to equilibrium is a sign of death. (29)

Compare this with the following statement, derived from the investigation of animal growth:

> Every organic form is the expression of a flux of processes. It persists only in a continuous change of its components. Every organic system appears stationary if considered from a certain point of view; but if we go a step deeper, we find that this maintenance involves continuous change of the systems of next lower order: of chemical compounds in the cell, of cells in multicellular organisms, of individuals in super-individual life units. It was said, in this sense ... that every organic system is essentially a hierarchical order of processes standing in dynamic equilibrium ... We may consider, therefore, organic forms as the expression of a pattern of processes of an ordered system of forces. This point of view can be called *dynamic morphology*. (5)

In fact, we have inferred, from quantitative analysis and theory of growth, and before the investigations of Schoenheimer and his co-workers, just the essential conclusions reached by the tracer method, namely, (a) that protein metabolism goes on, particularly in mammals, at much higher rates than classical physiology supposed, and (b) that there is synthesis and resynthesis of amino acids and proteins from ammonia and nitrogen-free chains (3). These predictions were rather hazardous at that time; but they have been fully confirmed by later isotope work, especially with N^{15}.

To apply the conception of open systems quantitatively to phenomena in the organism-as-a-whole, we have to use a sort of generalized kinetics. Since it is impossible to take into account the inextricable and largely unknown processes of intermediary metabolism, we use balance values for their statistical result. This procedure is in no way unusual. Already in chemistry, gross formulas – for example, those for photosynthesis or oxidation – indicate the net result of long chains of many partly unknown reaction steps. The same procedure is applied on a higher level in physiology when total metabolism is measured by oxygen consumption and carbon dioxide and calorie production, and bulk expressions, like Rubner's surface rule, are formulated; or when in clinical routine the diagnosis of, say, hyperthyroidism is based upon determination of basal metabolism. A similar procedure leads to exact theories of important biological phenomena.

Thus, a quantitative theory of growth has been developed. Growth is considered to be the result of the counteraction of anabolism and catabolism of the building materials. By quantitative expressions, using the physiological values of anabolism and catabolism and their size dependence, an explanation of growth in its general course, as well as in its details, and quantitative growth laws have been established. This theory is almost unique in physiology, for it permits precise quantitative predictions which have been verified, often in a very surprising way, by later experiments. The conceptions of dynamic morphology have been applied to a large number of problems, including the quantitative analysis of growth in micro-organisms, invertebrates, and vertebrates, the physiological connexions between metabolism and growth, leading to the establishment of metabolic types and corresponding growth types, allometry, growth gradients and physiological gradients, pharmacodynamic action, and phylogenetic problems (8).

Spiegelman (32) has given a quantitative theory of competition, regulation, dominance, and determination in morphogenesis, based upon a generalized kinetics of open systems and the gradient principle. The steady state and turnover rate of tissues are investigated by Leblond and his group (19a).

Excitation has already been considered by Hering a reversible disturbance of the processes going on in the living organism. The conception that organic systems are not in equilibrium, but in a steady state, has been advanced by Hill (17), mainly upon considerations of the osmotic nonequilibrium in living cells. Hill's theory of excitation (18), which is formally identical with that of Rashevsky (27), concerns a special case of steady states.

Under certain conditions, the approach in open systems to a steady state is not simply asymptotical, but shows an 'overshoot' or 'false start', as demonstrated by Burton (12), and Denbigh *et al.* (16). These phenomena are missed in ordinary physical chemistry, but are common in biological phenomena, e.g. in the sequence of afterpotentials following the spike potential in nerve excitation, in afterdischarge after inhibition, and in supernormal respiration after entering an oxygen debt.

Also the theory of feedback mechanisms (36), much discussed in the last few years, is related to the theory of open systems. Feedbacks, in man-made machines as well as in organisms, are based upon structural arrangements. Such mechanisms are present in the adult organism, and are responsible for homeostasis. However, the primary regulability, as manifested, for example, in embryonic regulations, and also in the nervous system after injuries, etc., is based upon direct dynamic interactions (9, 10).

In pharmacology, a conception corresponding to that of open systems has been applied by Loewe (21); quantitative relations have been derived for systems corresponding to the action of certain drugs ('put in', 'drop in', 'block out' systems). A deduction of the several laws of pharmacodynamic action from the organismic conception (von Bertalanffy) and a general system function was given by Werner (34).

Conceptions and systems of equations similar to those of open systems in physicochemistry and physiology appear in biocoenology, demography, and sociology (14, 19, 33).

The formal correspondence of general principles, irrespective of the kind of relations or forces between the components, leads to the conception of a 'General System Theory' as a new scientific doctrine, concerned with the principles which apply to systems in general.

Thus, the theory of open systems opens a new field in physics, and this development is even more remarkable because thermodynamics seemed to be a consummate doctrine within classical physics. In biology, the nature of the open system is at the basis of fundamental life phenomena, and this conception seems to point the direction and pave the way for biology to become an exact science.

References

1. B. BAVINK, *Ergebnisse und Probleme der Naturwissenschaften*, A. Francke, 1944.
2. L. VON BERTALANFFY, *Biologisches Zentralblatt*, vol. 49 (1929), p. 83.
3. L. VON BERTALANFFY, *Human Biology*, vol. 10 (1938), p. 181.
4. L. VON BERTALANFFY, *Naturwissenschaften*, vol. 28 (1940), p. 521.
5. L. VON BERTALANFFY, *Biologia Generalis*, vol. 15 (1941), p. 1.
6. L. VON BERTALANFFY, *Theoretische Biologie*, Gebruder Borntraeger, 1942.
7. L. VON BERTALANFFY, *Naturwissenschaften*, vol. 32 (1944), p. 26.
8. L. VON BERTALANFFY, *Experimentia*, vol. 4 (1948), p. 255; *Nature*, vol. 163 (1949), p. 156.
9. L. VON BERTALANFFY, *Biologia Generalis*, vol. 19 (1949), p. 114.
10. L. VON BERTALANFFY, *Das Biologische Weltbild*, A. Francke, 1949.
11. L. VON BERTALANFFY, *Nature*, vol. 163 (1949), p. 384.
12. A. C. BURTON, *Journal of Cellular and Comparative Physiology*, vol. 14 (1939), p. 327.
13. R. E. D. CLARK, *Darwin: Before and After*, Colin, 1948.
14. U. D'ANCONA, *Abh. z. exakten Biol.*, vol. 1 (1939).
15. U. DEHLINGER and E. WERTZ, *Naturwissenschaften*, vol. 30 (1942), p. 250.
16. K. G. DENBIGH, M. HICKS and F. M. PAGE, *Translations of the Faraday Society*, vol. 44 (1948), p. 479.
17. A. V. HILL, *Adventures in Biophysics*, University of Pennsylvania Press, 1931.
18. A. V. HILL, *Proceedings of the Royal Society*, vol. B11 (1936).
19. A. KOSTITZIN, *Biologie Mathématique*, Colin, 1937.
19a. C. P. LEBLOND and C. E. STEVENS, *Anatomical Record*, vol. 100 (1948), p. 357.
20. W. W. LEPESCHKIN, Protoplasma-Monographs, vol. 12 (1937); *Berichte Deutsche Botanische Gesellschaft*, vol. 57 (1939).
21. S. LOEWE, *Ergebnisse der Physiologie*, vol. 27 (1938), p. 47.

22. C. G. LONGWORTH, *Cold Spr. Harb. Sympos. quant. Biol.*, vol. 2 (1934), p. 213.
23. W. J. V. OSTERHOUT, *Journal of General Physiology*, vol. 16 (1933), p. 529.
24. W. J. V. OSTERHOUT and W. M. STANLEY, *Journal of General Physiology*, vol. 15 (1932), p. 667.
25. I. PRIGOGINE, *Etude Thermodynamique des Phénomènes Irréversibles*, Durrod, 1947.
26. I. PRIGOGINE, and J. M. WIAME, *Experientia*, vol. 2 (1946), p. 450.
27. N. RASHEVSKY, *Mathematical Biophysics*, Chicago University Press, 1938.
28. J. M. REINER and S. SPIEGELMAN, *Journal of Physical Chemistry*, vol. 49 (1945), p. 81.
29. D. RITTENBERG, *Journal of Mount Sinai Hospital*, vol. 14 (1948), p. 891.
30. P. VON RYSSELBERGHE, *Scientia*, vol. 83 (1948), p. 60.
31. A. SKRABAL, *Oesterreichische Chemiker-Zeitung*, vol. 48 (1947), p. 158; *Mh Chemistry*, vol. 80 (1949), p. 21.
32. S. SPIEGELMAN, *Quarterly Review of Biology*, vol. 20 (1945), p. 121.
33. V. VOLTERRA, *Leçons sur la Théorie Mathématique de la Tuite pour la Vie*, Gauthier-Villars, 1931.
34. G. WERNER, *Sitz. Akad. Wiss, Math.-naturw. Kl*, Abt. 11a, vol. 156 (1947).
35. G. WERNER and F. HOBBIGER, *Zeitschrift für Vitamin-Hormonund Fermentforschung*, vol. 2 (1948), p. 234; *Arch. int. Pharmacodyn.* vol. 79 (1949), p. 221.
36. N. WEINER, *Cybernetics*, Wiley, 1948; *Annals of the New York Academy of Science*, vol. 50 (1948), p. 187.
37. R. WOLTERECK, *Ontologie des Lebendigen*, F. Enke, 1940.

5 D. Katz and R. L. Kahn

Common Characteristics of Open Systems

D. Katz and R. L. Kahn, *The Social Psychology of Organizations*, chapter 2, Wiley, 1966, pp. 14–29.

The aims of social science with respect to human organizations are like those of any other science with respect to the events and phenomena of its domain. The social scientist wishes to understand human organizations, to describe what is essential in their form, aspects, and functions. He wishes to explain their cycles of growth and decline, to predict their effects and effectiveness. Perhaps he wishes as well to test and apply such knowledge by introducing purposeful changes into organizations – by making them, for example, more benign, more responsive to human needs.

Such efforts are not solely the prerogative of social science, however; common-sense approaches to understanding and altering organizations are ancient and perpetual. They tend, on the whole, to rely heavily on two assumptions: that the location and nature of an organization are given by its name; and that an organization is possessed of built-in goals – because such goals were implanted by its founders, decreed by its present leaders, or because they emerged mysteriously as the purposes of the organizational system itself. These assumptions scarcely provide an adequate basis for the study of organizations and at times can be misleading and even fallacious. We propose, however, to make use of the information to which they point.

The first problem in understanding an organization or a social system is its location and identification. How do we know that we are dealing with an organization? What are its boundaries? What behavior belongs to the organization and what behavior lies outside it? Who are the individuals whose actions are to be studied and what segments of their behavior are to be included?

The fact that popular names exist to label social organizations is both a help and a hindrance. These popular labels represent the socially accepted stereotypes about organizations and do not specify their role structure, their psychological nature, or their boundaries. On the other hand, these names help in locating the area of behavior in which we are interested. Moreover, the fact that people both within and without an organization accept stereotypes about its nature and functioning is one determinant of its character.

The second key characteristic of the common-sense approach to understanding an organization is to regard it simply as the epitome of the purposes of its designer, its leaders, or its key members. The teleology of this approach is again both a help and a hindrance. Since human purpose is deliberately built into organizations and is specifically recorded in the social compact, the by-laws, or other formal protocol of the undertaking, it would be inefficient not to utilize these sources of information. In the early development of a group, many processes are generated which have little to do with its rational purpose, but over time there is a cumulative recognition of the devices for ordering group life and a deliberate use of these devices.

Apart from formal protocol, the primary mission of an organization as perceived by its leaders furnishes a highly informative set of clues for the researcher seeking to study organizational functioning. Nevertheless, the stated purposes of an organization as given by its by-laws or in the reports of its leaders can be misleading. Such statements of objectives may idealize, rationalize, distort, omit, or even conceal some essential aspects of the functioning of the organization. Nor is there always agreement about the mission of the organization among its leaders and members. The university president may describe the purpose of his institution as one of turning out national leaders; the academic dean sees it as imparting the cultural heritage of the past, the academic vice-president as enabling students to move toward self-actualization and development, the graduate dean as creating new knowledge, the dean of men as training youngsters in technical and professional skills which will enable them to earn their living, and the editor of the student newspaper as inculcating the conservative values which will preserve the status quo of an outmoded capitalistic society.

The fallacy here is one of equating the purposes of goals of organizations with the purposes and goals of individual members. The organization as a system has an output, a product, or an outcome, but this is not necessarily identical with the individual purposes of group members. Though the founders of the organization and its key members do think in teleological terms about organizational objectives, we should not accept such practical thinking, useful as it may be, in place of a theoretical set of constructs for purposes of scientific analysis. Social science, too frequently in the past, has been misled by such short-cuts and has equated popular phenomenology with scientific explanation.

In fact, the classic body of theory and thinking about organizations has assumed a teleology of this sort as the easiest way of identifying organizational structures and their functions. From this point of view an organization is a social device for efficiently accomplishing through group means some stated purpose; it is the equivalent of the blueprint for the design of the machine which is to be created for some practical objective. The essential difficulty with this purposive or design approach is that an organization characteristically includes more and less than is indicated by the design of its founder or the purpose of its leader. Some of the factors assumed in the design may be lacking or so distorted in operational practice as to be meaningless, while unforeseen embellishments dominate the organizational structure. Moreover, it is not always possible to ferret out the designer of the organization or to discover the intricacies of the design which he carried in his head. The attempt by Merton (1957) to deal with the latent function of the organization in contrast with its manifest function is one way of dealing with this problem. The study of unanticipated consequences as well as anticipated consequences of organizational functioning is a similar way of handling the matter. Again, however, we are back to the purposes of the creator or leader, dealing with unanticipated consequences on the assumption that we can discover the consequences anticipated by him and can lump all other outcomes together as a kind of error variance.

It would be much better theoretically, however, to start with concepts which do not call for identifying the purposes of the designers and then correcting for them when they do not seem to be fulfilled. The theoretical concepts should begin with the input, output, and functioning of the organization as a system and not with the rational purposes of its leaders. We may want to utilize such purposive notions to lead us to sources of data or as subjects of special study, but not as our basic theoretical constructs for understanding organizations.

Our theoretical model for the understanding of organizations is that of an energic input–output system in which the energic return from the output reactivates the system. Social organizations are flagrantly open systems in that the input of energies and the conversion of output into further energic input consist of transactions between the organization and its environment.

All social systems, including organizations, consist of the patterned activities of a number of individuals. Moreover, these patterned activities are complementary or interdependent with respect to some common output or outcome; they are repeated, relatively enduring, and bounded in space and time. If the activity pattern occurs only once or at unpredictable intervals, we could not speak of an organization. The stability or recurrence of activities can be examined in relation to the *energic input* into the system, the *transformation of energies within the system*, and the *resulting product or energic output*. In a factory the raw materials and the human labor are the energic input, the patterned activities of production the transformation of energy, and the finished product the output. To maintain this patterned activity requires a continued renewal of the inflow of energy. This is guaranteed in social systems by the energic return from the product or outcome. Thus the outcome of the cycle of activities furnishes new energy for the initiation of a renewed cycle. The company which produces automobiles sells them and by doing so obtains the means of securing new raw materials, compensating its labor force, and continuing the activity pattern.

In many organizations outcomes are converted into money and new energy is furnished through this mechanism. Money is a convenient way of handling energy units both on the output and input sides, and buying and selling represent one set of social rules for regulating the exchange of money. Indeed, these rules are so effective and so widespread that there is some danger of mistaking the business of buying and selling for the defining cycles of organization. It is a commonplace executive observation that businesses exist to make money, and the observation is usually allowed to go unchallenged. It is, however, a very limited statement about the purposes of business.

Some human organizations do not depend on the cycle of selling and buying to maintain themselves. Universities and public agencies depend rather on bequests and legislative appropriations, and in so-called voluntary organizations the output re-energizes the activity of organization members in a more direct fashion. Member activities and accomplishments are rewarding in themselves and tend therefore to be continued, without the mediation of the outside environment. A society of bird watchers can wander into the hills and engage in the rewarding activities of identifying birds for their mutual edification and enjoyment. Organizations thus differ on this important dimension of the source of energy renewal, with the great majority utilizing both intrinsic and extrinsic sources in varying degree. Most large-scale organizations are not as self-contained as small voluntary groups and are very dependent upon the social effects of their output for energy renewal.

Our two basic criteria for identifying social systems and determining their functions are (1) tracing the pattern of energy exchange or activity of people as it results in some output and (2) ascertaining how the output is translated into energy which reactivates the pattern. We shall refer to organizational functions or objectives not as the conscious purposes of group leaders or group members but as the outcomes which are the energic source for a maintenance of the same type of output.

This model of an energic input–output system is taken from the open system theory as promulgated by von Bertalanffy (1956). Theorists have pointed out the applicability of the system concepts of the natural sciences to the problems of social science. It is important, therefore, to examine in more detail the constructs of system theory and the characteristics of open systems.

System theory is basically concerned with problems of relationships, of structure, and of interdependence rather than with the constant attributes of objects. In general approach it resembles field theory except that its dynamics deal with temporal as well as spatial patterns. Older formulations of system constructs dealt with the closed systems of the physical sciences, in which relatively self-contained structures could be treated successfully as if they were independent of external forces. But living systems, whether biological organisms or social organizations, are acutely dependent upon their external environment and so must be conceived of as open systems.

Before the advent of open-system thinking, social scientists tended to take one of two approaches in dealing with social structures; they tended either (1) to regard them as closed systems to which the laws of physics applied or (2) to endow them with some vitalistic concept like entelechy. In the former case they ignored the environmental forces affecting the organization and in the latter case they fell back upon some magical purposiveness to account for organizational functioning. Biological theorists, however, have rescued us from this trap by pointing out that the concept of the open system means that we neither have to follow the laws of traditional physics, nor in deserting them do we have to abandon science. The laws of Newtonian physics are correct generalizations but they are limited to closed systems. They do not apply in the same fashion to open systems which maintain themselves through constant commerce with their environment, i.e. a continuous inflow and outflow of energy through permeable boundaries.

One example of the operation of closed versus open systems can be seen in the concept of entropy and the second law of thermodynamics. According to the second law of thermodynamics, a system moves toward equilibrium; it tends to run down, that is, its differentiated structures tend to move toward dissolution as the elements composing them become arranged in random disorder. For example, suppose that a bar of iron has been heated by the application of a blowtorch on one side. The arrangement of all the fast (heated) molecules on one side and all the slow molecules on the other is an unstable state, and over time the distribution of molecules becomes in effect random, with the resultant cooling of one side and heating of the other, so that all surfaces of the iron approach the same temperature. A similar process of heat exchange will also be going on between the iron bar and its environment, so that the bar will gradually approach the temperature of the room in which it is located, and in doing will elevate somewhat the previous temperature of the room. More technically, entropy increases toward a maximum and equilibrium occurs as the physical system attains the state of the most probable distribution of its elements. In social systems, however, structures tend to become more elaborated rather than less differentiated. The rich may grow richer and the poor may grow poorer. The open system does not run down, because it can import energy from the world around it. Thus the operation of entropy is counteracted by the importation of energy and the living system is characterized by negative rather than positive entropy.

Common Characteristics of Open Systems

Though the various types of open systems have common characteristics by virtue of being open systems, they differ in other characteristics. If this were not the case, we would be able to obtain all our basic knowledge about social organizations through studying the biological organisms or even through the study of a single cell.

The following nine characteristics seem to define all open systems:

1. Importation of energy

Open systems import some form of energy from the external environment. The cell receives oxygen from the blood stream; the body similarly takes in oxygen from the air and food from the external world. The personality is dependent upon the external world for stimulation. Studies of sensory deprivation show that when a person is placed in a darkened soundproof room, where he has a minimal amount of visual and auditory stimulation, he develops hallucinations and other signs of mental stress (Solomon *et al.*, 1961). Deprivation of social stimulation also can lead to mental disorganization (Spitz, 1945). Köhler's (1944, 1947) studies of the figural after-effects of continued stimulation show the dependence of perception upon its energic support from the external world. Animals deprived of visual experience from birth for a prolonged period never fully recover their visual capacities (Melzack and Thompson, 1956). In other words, the functioning personality is heavily dependent upon the continuous inflow of stimulation from the external environment. Similarly, social organizations must also draw renewed supplies of energy from other institutions, or people, or the material environment. No social structure is self-sufficient or self-contained.

2. The through-put

Open systems transform the energy available to them. The body converts starch and sugar into heat and action. The personality converts chemical and electrical forms of stimulation into sensory qualities, and information into thought patterns. The organization creates a new product, or processes materials, or trains people, or provides a service. These activities entail some reorganization of input. Some work gets done in the system.

3. The output

Open systems export some product into the environment, whether it be the invention of an inquiring mind or a bridge constructed by an engineering firm. Even the biological organism exports physiological products such as carbon dioxide from the lungs which helps to maintain plants in the immediate environment.

4. Systems as cycles of events

The pattern of activities of the energy exchange has a cyclic character. The product exported into the environment furnishes the sources of energy for the repetition of the cycle of activities. The energy reinforcing the cycle of activities can derive from some exchange of the product in the external world or from the activity itself. In the former instance, the industrial concern utilizes raw materials and human labor to turn out a product which is marketed, and the monetary return is used to obtain more raw materials and labor to perpetuate the cycle of activities. In the latter instance, the voluntary organization can provide expressive satisfactions to its members so that the energy renewal comes directly from the organizational activity itself.

The problem of structure, or the relatedness of parts, can be observed directly in some physical arrangement of things where the larger unit is physically bounded and its subparts are also bounded within the larger structure. But how do we deal with social structures, where physical boundaries in this sense do not exist? It was the genius of F. H. Allport (1962) which contributed the answer, namely that the structure is to be found in an inter-related set of events which return upon themselves to complete and renew a cycle of activities. It is events rather than things which are structured, so that social structure is a dynamic rather than a static concept. Activities are structured so that they comprise a unity in their completion or closure. A simple linear stimulus–response exchange between two people would not constitute social structure. To create structure, the responses of A would have to elicit B's reactions in such a manner that the responses of the latter would stimulate A to further responses. Of course the chain of events may involve many people, but their behavior can be characterized as showing structure only when there is some closure to the chain by a return to its point of origin with the probability that the chain of events will then be repeated. The repetition of the cycle does not have to involve the same set of phenotypical happenings. It may expand to include more sub-events of exactly the same kind or it may involve similar activities directed toward the same outcomes. In the individual organism the eye may move in such a way as to have the point of light fall upon the center of the retina. As the point of light moves, the movements of the eye may also change but to complete the same cycle of activity, i.e. to focus upon the point of light.

A single cycle of events of a self-closing character gives us a simple form of structure. But such single cycles can also combine to give a larger structure of events or an event system. An event system may consist of a circle of smaller cycles or hoops, each one of which makes contact with several others. Cycles may also be tangential to one another from other types of subsystems. The basic method for the identification of social structures is to follow the energic chain of events from the input of energy through its transformation to the point of closure of the cycle.

5. Negative entropy

To survive, open systems must move to arrest the entropic process; they must acquire negative entropy. The entropic process is a universal law of nature in which all forms of organization move toward disorganization or death. Complex physical systems move toward simple random distribution of their elements and biological organisms also run down and perish. The open system, however, by importing more energy from its environment than it expends, can store energy and can acquire negative entropy. There is then a general trend in an open system to maximize its ratio of imported to expended energy, to survive and even during periods of crisis to live on borrowed time. Prisoners in concentration camps on a starvation diet will carefully conserve any form of energy expenditure to make the limited food intake go as far as possible (Cohen, 1954). Social organizations will seek to improve their survival position and to acquire in their reserves a comfortable margin of operation.

The entropic process asserts itself in all biological systems as well as in closed physical systems. The energy replenishment of the biological organism is not of a qualitative character which can maintain indefinitely the complex organizational structure of living tissue. Social systems, however, are not anchored in the same physical constancies as biological organisms and so are capable of almost indefinite arresting of the entropic process. Nevertheless the number of organizations which go out of existence every year is large.

6. Information input, negative feedback, and the coding process

The inputs into living systems consist not only of energic materials which become transformed or altered in the work that gets done. Inputs are also informative in character and furnish signals to the structure about the environment and about its own functioning in relation to the environment. Just as we recognize the distinction between cues and drives in individual psychology, so must we take account of information and energic inputs for all living systems.

The simplest type of information input found in all systems is negative feedback. Information feedback of a negative kind enables the system to correct its deviations from course. The working parts of the machine feed back information about the effects of their operation to some central mechanism or subsystem which acts on such information to keep the system on target. The thermostat which controls the temperature of the room is a simple example of a regulatory device which operates on the basis of negative feedback. The automated power plant would furnish more complex examples. Miller (1955) emphasizes the critical nature of negative feedback in his proposition: '*When a system's negative feedback discontinues, its steady state vanishes, and at the same time its boundary disappears and the system terminates*' (p. 529). If there is no corrective device to get the system back on its course, it will expend too much energy or it will ingest too much energic input and no longer continue as a system.

The reception of inputs into a system is selective. Not all energic inputs are capable of being absorbed into every system. The digestive system of living creatures assimilates only those inputs to which it is adapted. Similarly, systems can react only to those information signals to which they are attuned. The general term for the selective mechanisms of a system by which incoming materials are rejected or accepted and translated for the structure is coding. Through the coding process the 'blooming, buzzing confusion' of the world is simplified into a few meaningful and simplified categories for a given system. The nature of the functions performed by the system determines its coding mechanisms, which in turn perpetuate this type of functioning.

7. The steady state and dynamic homeostasis

The importation of energy to arrest entropy operates to maintain some constancy in energy exchange, so that open systems which survive are characterized by a steady state. A steady state is not motionless or a true equilibrium. There is a continuous inflow of energy from the external environment and a continuous export of the products of the system, but the character of the system, the ratio of the energy exchanges and the relations between parts, remains the same. The catabolic and anabolic processes of tissue breakdown and restoration within the body preserve a steady state so that the organism from time to time is not the identical organism it was but a highly similar organism. The steady state is seen in clear form in the homeostatic processes for the regulation of body temperature; external conditions of humidity and temperature may vary, but the temperature of the body remains the same. The endocrine glands are a regulatory mechanism for preserving an evenness of physiological functioning. The general principle here is that of Le Chatelier (see Bradley and Calvin,

1956) who maintains that any internal or external factor making for disruption of the system is countered by forces which restore the system as closely as possible to its previous state. Krech and Crutchfield (1948) similarly hold, with respect to psychological organization, that cognitive structures will react to influences in such a way as to absorb them with minimal change to existing cognitive integration.

The homeostatic principle does not apply literally to the functioning of all complex living systems, in that in counteracting entropy they move toward growth and expansion. This apparent contradiction can be resolved, however, if we recognize the complexity of the subsystems and their interaction in anticipating changes necessary for the maintenance of an overall steady state. Stagner (1951) has pointed out that the initial disturbance of a given tissue constancy within the biological organism will result in mobilization of energy to restore the balance, but that recurrent upsets will lead to actions to anticipate the disturbance:

> We eat before we experience intense hunger pangs . . . energy mobilization for forestalling tactics must be explained in terms of a *cortical tension* which reflects the visceral-proprioceptic pattern of the original biological disequilibration . . . *Dynamic homeostasis* involves the maintenance of tissue constancies by establishing a constant physical environment – by reducing the variability and disturbing effects of external stimulation. Thus the organism does not simply restore the prior equilibrium. A new, more complex and more comprehensive equilibrium is established (p. 5).

Though the tendency toward a steady state in its simplest form is homeostatic, as in the preservation of a constant body temperature, the basic principle is *the preservation of the character of the system*. The equilibrium which complex systems approach is often that of a quasi-stationary equilibrium, to use Lewin's concept (1947). An adjustment in one direction is countered by a movement in the opposite direction and both movements are approximate rather than precise in their compensatory nature. Thus a temporal chart of activity will show a series of ups and downs rather than a smooth curve.

In preserving the character of the system, moreover, the structure will tend to import more energy than is required for its output, as we have already noted in discussing negative entropy. To insure survival, systems will operate to acquire some margin of safety beyond the immediate level of existence. The body will store fat, the social organization will build up reserves, the society will increase its technological and cultural base. Miller (1955) has formulated the proposition that the rate of growth of a system – within certain ranges – is exponential if it exists in a medium which makes available unrestricted amounts of energy for input.

In adapting to their environment, systems will attempt to cope with external forces by ingesting them or acquiring control over them. The physical boundedness of the single organism means that such attempts at control over the environment affect the behavioral system rather than the biological system of the individual. Social systems will move, however, toward incorporating within their boundaries the external resources essential to survival. Again the result is an expansion of the original system.

Thus, the steady state which at the simple level is one of homeostasis over time, at more complex levels becomes one of preserving the character of the system through growth and expansion. The basic type of system does not change directly as a consequence of expansion. The most common type of growth is a multiplication of the same type of cycles or subsystems – a change in quantity rather than in quality. Animals and plant species grow by multiplication. A social system adds more units of the same essential type as it already has. Haire (1959) has studied the ratio between the sizes of different subsystems in growing business organizations. He found that though the number of people increased in both the production subsystem and the subsystem concerned with the external world, the ratio of the two groups remained constant. Qualitative change does occur, however, in two ways. In the first place, quantitative growth calls for supportive subsystems of a specialized character not necessary when the system was smaller. In the second place, there is a point where quantitative changes produce a qualitative difference in the functioning of a system. A small college which triples its size is no longer the same institution in terms of the relation between its administration and faculty, relations among the various academic departments, or the nature of its instruction.

In fine, living systems exhibit a growth or expansion dynamic in which they maximize their basic character. They react to change or they anticipate change through growth which assimilates the new energic inputs to the nature of their structure. In terms of Lewin's quasi-stationary equilibrium the ups and downs of the adjustive process do not always result in a return to the old level. Under certain circumstances a solidification or freezing occurs during one of the adjustive cycles. A new base line level is thus established and successive movements fluctuate around this plateau which may be either above or below the previous plateau of operation.

8. *Differentiation*

Open systems move in the direction of differentiation and elaboration. Diffuse global patterns are replaced by more specialized functions. The sense organs and the nervous system evolved as highly differentiated structures from the primitive nervous tissues. The growth of the personality proceeds from primitive, crude organizations of mental functions to hierarchically structured and well-differentiated systems of beliefs and feelings. Social organizations move toward the multiplication and elaboration of roles with greater specialization of function. In the United States today medical specialists now outnumber the general practitioners.

One type of differentiated growth in systems is what von Bertalanffy (1956) terms progressive mechanization. It finds expression in the way in which a system achieves a steady state. The early method is a process which involves an interaction of various dynamic forces, whereas the later development entails the use of a regulatory feedback mechanism. He writes:

> It can be shown that the *primary* regulations in organic systems, that is, those which are most fundamental and primitive in embryonic development as well as in evolution, are of such nature of dynamic interaction ... Superimposed are those regulations which we may call *secondary*, and which are controlled by fixed arrangements, especially of the feedback type. This state of affairs is a consequence of a general principle of organization which may be called progressive mechanization. At first, systems – biological, neurological, psychological or social – are governed by dynamic interaction of their components; later on, fixed arrangements and conditions of constraint are established which

render the system and its parts more efficient, but also gradually diminish and eventually abolish its equipotentiality (p. 6).

9. Equifinality

Open systems are further characterized by the principle of equifinality, a principle suggested by von Bertalanffy in 1940. According to this principle, a system can reach the same final state from differing initial conditions and by a variety of paths. The well-known biological experiments on the sea urchin show that a normal creature of that species can develop from a complete ovum, from each half of a divided ovum, or from the fusion product of two whole ova. As open systems move toward regulatory mechanisms to control their operations, the amount of equifinality may be reduced.

Some Consequences of Viewing Organizations as Open Systems

In the following chapter we shall inquire into the specific implications of considering organizations as open systems and into the ways in which social organizations differ from other types of living systems. At this point, however, we should call attention to some of the misconceptions which arise both in theory and practice when social organizations are regarded as closed rather than open systems.

The major misconception is the failure to recognize fully that the organization is continually dependent upon inputs from the environment and that the inflow of materials and human energy is not a constant. The fact that organizations have built-in protective devices to maintain stability and that they are notoriously difficult to change in the direction of some reformer's desires should not obscure the realities of the dynamic interrelationships of any social structure with its social and natural environment. The very efforts of the organization to maintain a constant external environment produce changes in organizational structure. The reaction to changed inputs to mute their possible revolutionary implications also results in changes.

The typical models in organizational theorizing concentrate upon principles of internal functioning as if these problems were independent of changes in the environment and as if they did not affect the maintenance inputs of motivation and morale. Moves toward tighter integration and coordination are made to insure stability, when flexibility may be the more important requirement. Moreover, coordination and control become ends in themselves rather than means to an end. They are not seen in full perspective as adjusting the system to its environment but as desirable goals within a closed system. In fact, however, every attempt at co-ordination which is not functionally required may produce a host of new organizational problems.

One error which stems from this kind of misconception is the failure to recognize the equifinality of the open system, namely that there are more ways than one of producing a given outcome. In a closed physical system the same initial conditions must lead to the same final result. In open systems this is not true even at the biological level. It is much less true at the social level. Yet in practice we insist that there is one best way of assembling a gun for all recruits, one best way for the baseball player to hurl the ball in from the outfield, and that we standardize and teach these best methods. Now it is true under certain conditions that there is one best way, but these conditions must first be established. The general principle, which characterizes all open systems, is that there does not have to be a single method for achieving an objective.

A second error lies in the notion that irregularities in the functioning of a system due to environmental influences are error variances and should be treated accordingly. According to this conception, they should be controlled out of studies of organizations. From the organization's own operations they should be excluded as irrelevant and should be guarded against. The decisions of officers to omit a consideration of external factors or to guard against such influences in a defensive fashion, as if they would go away if ignored, is an instance of this type of thinking. So is the now outmoded 'public be damned' attitude of businessmen toward the clientele upon whose support they depend. Open system theory, on the other hand, would maintain that environmental influences are not sources of error variance but are integrally related to the functioning of a social system, and that we cannot understand a system without a constant study of the forces that impinge upon it.

Thinking of the organization as a closed system, moreover, results in a failure to develop the intelligence or feedback function of obtaining adequate information about the changes in environmental forces. It is remarkable how weak many industrial companies are in their market research departments when they are so dependent upon the market. The prediction can be hazarded that organizations in our society will increasingly move toward the improvement of the facilities for research in assessing environmental forces. The reason is that we are in the process of correcting our misconception of the organization as a closed system.

Emery and Trist (1960) have pointed out how current theorizing on organizations still reflects the older closed system conceptions. They write:

> In the realm of social theory, however, there has been something of a tendency to continue thinking in terms of a 'closed' system, that is, to regard the enterprise as sufficiently independent to allow most of its problems to be analysed with reference to its internal structure and without reference to its external environment . . . In practice the system theorists in social science . . . did 'tend to focus on the statics of social structure and to neglect the study of structural change'. In an attempt to overcome this bias, Merton suggested that 'the concept of dysfunction, which implied the concept of strain, stress and tension on the structural level, provides an analytical approach to the study of dynamics and change'. This concept has been widely accepted by system theorists but while it draws attention to sources of imbalance within an organization it does not conceptually reflect the mutual permeation of an organization and its environment that is the cause of such imbalance. It still retains the limiting perspectives of 'closed system' theorizing. In the administrative field the same limitations may be seen in the otherwise invaluable contributions of Barnard and related writers (p. 84).

Summary

The open-system approach to organizations is contrasted with common-sense approaches, which tend to accept popular names and stereotypes as basic organizational properties and to identify the purpose of an organization in terms of the goals of its founders and leaders.

The open-system approach, on the other hand, begins by identifying and mapping the repeated cycles of input, transformation, output, and renewed input which comprise the organizational pattern. This approach to organizations represents the adaptation of work in biology and in the physical sciences by von Bertalanffy and others.

Organizations as a special class of open systems have properties of their own, but they share other properties in common with all open systems. These include the importation of energy from the environment, the through-put or transformation of the imported energy into some product form which is characteristic of the system, the exporting of that product into the environment, and the re-energizing of the system from sources in the environment.

Open systems also share the characteristics of negative entropy, feedback, homeostasis, differentiation, and equifinality. The law of negative entropy states that systems survive and maintain their characteristic internal order only so long as they import from the environment more energy than they expend in the process of transformation and exportation. The feedback principle has to do with information input, which is a special kind of energic importation, a kind of signal to the system about environmental conditions and about the functioning of the system in relation to its environment. The feedback of such information enables the system to correct for its own malfunctioning or for changes in the environment, and thus to maintain a steady state or homeostasis. This is a dynamic rather than a static balance, however. Open systems are not at rest but tend toward differentiation and elaboration, both because of subsystem dynamics and because of the relationship between growth and survival. Finally, open systems are characterized by the principle of equifinality, which asserts that systems can reach the same final state from different initial conditions and by different paths of development.

Traditional organizational theories have tended to view the human organization as a closed system. This tendency has led to a disregard of differing organizational environments and the nature of organizational dependency on environment. It has led also to an overconcentration on principles of internal organizational functioning, with consequent failure to develop and understand the processes of feedback which are essential to survival.

References

ALLPORT, F. H. (1962), 'A structuronomic conception of behavior: individual and collective. I. Structural theory and the master problem of social psychology', *Journal of Abnormal and Social Psychology*, vol. 64, pp. 3–30.

BRADLEY, D. F., and CALVIN, M. (1956), 'Behavior: imbalance in a network of chemical transformations', *General Systems*, Yearbook of the Society for the Advancement of General System Theory, vol. 1, pp. 56–65.

COHEN, E. (1954), *Human Behavior in the Concentration Camp*, Jonathan Cape.

EMERY, F. E., and TRIST, E. L. (1960), 'Socio-technical systems', in *Management Sciences Models and Techniques*, vol. 2, Pergamon Press.

HAIRE, M. (1959), 'Biological models and empirical histories of the growth of organizations', in M. Haire (ed.), *Modern Organization Theory*, Wiley, pp. 272–306.

KOEHLER, W., and EMERY, D. (1947), 'Figural after-effects in the third dimension of visual space', *American Journal of Psychology*, vol. 60, pp. 159–201.

KOEHLER, W., and WALLACH, H. (1944), 'Figural after-effects: an investigation of visual processes', *Proceedings of the American Philosophical Society*, vol. 88, pp. 269–357.

KRECH, D., and CRUTCHFIELD, R. (1948), *Theory and Problems of Social Psychology*, McGraw-Hill.

LEWIN, K. (1947), 'Frontiers in group dynamics', *Human Relations*, vol. 1, pp. 5–41.

MELZACK, R., and THOMPSON, W. (1956), 'Effects of early experience on social behaviour', *Canadian Journal of Psychology*, vol. 10. pp. 82–90.

MERTON, R. K. (1957), *Social Theory and Social Structure*, Free Press, rev. edn.

MILLER, J. G. (1955), 'Toward a general theory for the behavioral sciences', *American Psychologist*, vol. 10. pp. 513–31.

SOLOMON, P. *et al.* (1961), *Sensory Deprivation*, Harvard University Press.

SPITZ, R. A. (1945), 'Hospitalism: an inquiry into the genesis of psychiatric conditions in early childhood', *Psychoanalytic Study of the Child*, vol. 1. pp. 53–74.

STAGNER, R. (1951), 'Homeostasis as a unifying concept in personality theory', *Psychological Review*, vol. 58. pp. 5–17.

VON BERTALANFFY, L. (1940), 'Der organismus als physikalisches system betrachtet', *Naturwissenschaften*, vol. 28. pp. 521ff.

VON BERTALANFFY, L. (1956), 'General system theory', *General Systems*, Yearbook of the Society for the Advancement of General System Theory, vol. 1. pp. 1–10.

6 W. R. Ashby

Self-regulation and Requisite Variety

W. R. Ashby, *Introduction to Cybernetics*, chapter 11, Wiley, 1956, pp. 202–18.

1.1. In the previous chapter [not included here] we considered regulation from the biological point of view, taking it as something sufficiently well understood. In this chapter we shall examine the process of regulation itself, with the aim of finding out exactly what is involved and implied. In particular we shall develop ways of *measuring* the amount or degree of regulation achieved, and we shall show that this amount has an upper limit.

1.2. The subject of regulation is very wide in its applications, covering as it does most of the activities in physiology, sociology, ecology, economics, and much of the activities in almost every branch of science and life. Further, the types of regulator that exist are almost bewildering in their variety. One way of treating the subject would be to deal seriatim with the various types; and Chapter 12 [not included here] will, in fact, indicate them. In this chapter, however, we shall be attempting to get at the core of the subject – to find what is common to all.

What is common to all regulators, however, is not, at first sight, much like any particular form. We will therefore start anew in the next section, making no explicit reference to what has gone before. Only after the new subject has been sufficiently developed will we begin to consider any relation it may have to regulation.

1.3. *Play and outcome.* Let us therefore forget all about regulation and simply suppose that we are watching two players, *R* and *D*, who are engaged in a game. We shall follow the fortunes of *R*, who is attempting to score an *a*. The rules are as follows. They have before them Table 1, which can be seen by both.

D must play first, by selecting a number, and thus a particular row. *R*, knowing this number, then selects a Greek letter, and thus a particular column. The italic letter specified by the intersection of the row and column is the *outcome*. If it is an *a*, *R* wins; if not, *R* loses.

Table 1

		R		
		α	β	γ
	1	*b*	*a*	*c*
D	2	*a*	*c*	*b*
	3	*c*	*b*	*a*

Examination of the table soon shows that with this particular table *R* can win always. Whatever value *D* selects first, *R* can always select a Greek letter that will give the desired outcome. Thus if *D* selects 1, *R* selects β; if *D* selects 2, *R* selects α; and so on. In fact, if *R* acts according to the transformation

$$\downarrow \begin{matrix} 1 & 2 & 3 \\ \beta & \alpha & \gamma \end{matrix}$$

then he can always force the outcome to be *a*.

R's position, with this particular table, is peculiarly favorable, for not only can *R* always force *a* as the outcome, but he can as readily force, if desired, *b* or *c* as the outcome. *R* has, in fact, complete control of the outcome.

Exercise 1. What transformation should *R* use to force *c* as outcome?
Exercise 2. If both *R*'s and *D*'s values are integers, and the outcome *E* is also an integer, given by

$$E = R - 2D$$

find an expression to give *R* in terms of *D* when the desired outcome is 37.
Exercise 3. A car's back wheels are skidding. *D* is the variable 'Side to which the tail is moving', with two values, Right and Left. *R* is the driver's action 'Direction in which he turns the steering wheel', with two values, Right and Left. Form the 2 × 2 table and fill in the outcomes.

Exercise 4. If *R*'s play is determined by *D*'s in accordance with the transformation

$$\downarrow \begin{matrix} 1 & 2 & 3 \\ \gamma & \beta & \alpha \end{matrix}$$

and many games are observed, what will be the variety in the many outcomes?
Exercise 5. Has *R* complete control of the outcome if the table is triunique?

1.4. The table used above is, of course, peculiarly favorable to *R*. Other tables are, however, possible. Thus suppose *D* and *R*, playing on the same rules, are now given Table 2 in which *D* now has a choice of five, and *R* a choice of four moves.

Table 2

		R			
		α	β	γ	δ
	1	*b*	*d*	*a*	*a*
	2	*a*	*d*	*a*	*d*
D	3	*d*	*a*	*a*	*a*
	4	*d*	*b*	*a*	*b*
	5	*d*	*a*	*b*	*d*

If *a* is the target, *R* can always win. In fact, if *D* selects 3, *R* has several ways of winning. As every row has at least one *a*, *R* can always force the appearance of *a* as the outcome. On the other hand, if the target is *b* he cannot always win. For if *D* selects 3, there is no move by *R* that will give *b* as the outcome. And if the target is *c*, *R* is quite helpless, for *D* wins always.

It will be seen that different arrangements within the table, and different numbers of states available to *D* and *R*, can give rise to a variety of situations from the point of view of *R*.

Exercise 1. With Table 1, can *R* always win if the target is *d*?
Exercise 2. (Continued.) What transformation should *R* use?
Exercise 3. (Continued.) If *a* is the target and *D*, for some reason, never plays 5, how can *R* simplify his method of play?
Exercise 4. A guest is coming to dinner, but the butler does not know who. He knows only that it may be Mr A, who drinks only sherry or

wine, Mrs B, who drinks only gin or brandy, or Mr C, who drinks only red wine, brandy, or sherry. In the cellar he finds he has only whisky, gin, and sherry. Can he find something acceptable to the guest, whoever comes?

1.5. Can any *general* statement be made about *R*'s modes of play and prospects of success?

If full generality is allowed in Table 2, the possibilities are so many, arbitrary and complicated that little can be said. There is one type, however, that allows a precise statement and is at the same time sufficiently general to be of interest. (It is also fundamental in the theory of regulation.)

Table 3

		R		
		α	β	γ
	1	*f*	*f*	*k*
	2	*k*	*e*	*f*
	3	*m*	*k*	*a*
	4	*b*	*b*	*b*
D	5	*c*	*q*	*c*
	6	*h*	*h*	*m*
	7	*j*	*d*	*d*
	8	*a*	*p*	*j*
	9	*l*	*n*	*h*

From all possible tables let us eliminate those that make *R*'s game too easy to be of interest. *Exercise 3* showed that if a column contains repetitions, *R*'s play need not be discriminating; that is, *R* need not change his move with each change of *D*'s move. Let us consider, then, only those tables in which *no column contains a repeated outcome*. When this is so *R* must select his move on *full* knowledge of *D*'s move; i.e. any change of *D*'s move must require a change on *R*'s part. (Nothing is assumed here about how the outcomes in one column are related to those in another, so these relations are unrestricted.) Such a table is Table 3. Now, some target being given, let *R* specify what his move will be for each move by *D*. What is essential is that, win or lose, he must specify one and only one move in response to each possible move of *D*. His specification, or 'strategy' as it might be called, might appear:

If *D* selects 1, I shall select γ
" " 2, " " α
" " 3, " " β
. . . . . .
" " 9, " " α

He is, of course, specifying a transformation (which must be single-valued, as *R* may not make two moves simultaneously):

$$\downarrow \begin{matrix} 1 & 2 & 3 & \ldots & 9 \\ \gamma & \alpha & \beta & \ldots & \alpha \end{matrix}$$

This transformation uniquely specifies a set of outcomes – those that will actually occur if *D*, over a sequence of plays, includes every possible move at least once. For 1 and γ give the outcome *k*, and so on, leading to the transformation:

$$\downarrow \begin{matrix} (1,\gamma) & (2,\alpha) & (3,\beta) & \ldots & (9,\alpha) \\ k & k & k & \ldots & l \end{matrix}$$

It can now be stated that the variety in this set of outcomes cannot be less than

$$\frac{D\text{'s variety}}{R\text{'s variety}}$$

i.e. in this case, 9/3.

It is easily proved. Suppose *R* marks one element in each row and concentrates simply on keeping the variety of the marked elements as small as possible (ignoring for the moment any idea of a target). He marks an element in the first row. In the second row he must change to a new column if he is not to increase the variety by adding a new, different, element; for in the initially selected column the elements are all different, by hypothesis. To keep the variety down to one element he must change to a new column at each row. (This is the *best* he can do; it may be that change from column to column is not sufficient to keep the variety down to one element, but this is irrelevant, for we are interested only in what is the least possible variety, assuming that everything falls as favorably as possible.) So if *R* has *n* moves available (three in the example), at the *n*th row all the columns are used, so one of the columns must be used again for the next row, and a new outcome *must* be allowed into the set of outcomes. Thus in Table 3, selection of the k's in the first three rows will enable the variety to be kept to one element, but at the fourth row a second element *must* be allowed into the set of outcomes.

In general, if no two elements in the same column are equal, and if a set of outcomes is selected by R, one from each row, and if the table has r rows and c columns, then *the variety in the selected set of outcomes cannot be fewer than* r/c.

The Law of Requisite Variety

1.6. We can now look at this game (still with the restriction that no element may be repeated in a column) from a slightly different point of view. If R's move is unvarying, so that he produces the same move, whatever D's move, then *the variety in the outcomes will be as large as the variety in D's moves. D* now is, as it were, exerting full control over the outcomes.

If next R uses, or has available, two moves, then the variety of the outcomes can be reduced to a half (but not lower). If R has three moves, it can be reduced to a third (but not lower); and so on. Thus if the variety in the outcomes is to be reduced to some assigned number, or assigned fraction of D's variety, R's variety *must* be increased to at least the appropriate minimum. *Only variety in R's moves can force down the variety in the outcomes.*

1.7. If the varieties are measured logarithmically (as is almost always convenient), and if the same conditions hold, then the theorem takes a very simple form. Let V_D be the variety of D, V_R that of R, and V_O that of the outcome (all measured logarithmically). Then the previous section has proved that V_O cannot be less, numerically, than the value of $V_D - V_R$. Thus V_O's minimum is $V_D - V_R$.

If V_D is given and fixed, $V_D - V_R$ can be lessened only by a corresponding increase in V_R. Thus *the variety in the outcomes, if minimal, can be decreased further only by a corresponding increase in that of R.* (A more general statement is given in 1.9.)

This is the Law of Requisite Variety. To put it more picturesquely: *only variety in R can force down the variety due to D; only variety can destroy variety.*

This thesis is so fundamental in the general theory of regulation that I shall give some further illustrations and proofs before turning to consider its actual application.

1.8. (This section can be omitted at first reading.) The law is of very general applicability, and by no means just a trivial outcome of the tabular form. To show that this is so, what is essentially the same theorem will be proved in the case when the variety is spread out in time and the fluctuation incessant – the case specially considered by Shannon. (The notation and concepts in this section are those of Shannon's book.)

Let D, R, and E be three variables, such that each is an information source, though 'source' here is not to imply that they are acting independently. Without any regard for how they are related causally, a variety of entropies can be calculated, or measured empirically. There is $H(D, R, E)$, the entropy of the vector that has the three as components; there is $H_D(E)$, the uncertainty in E when D's state is known; there is $H_{ED}(R)$, the uncertainty in R when both E and D are known; and so on.

The condition introduced in 1.5 (that no element shall occur twice in a column) here corresponds to the condition that if R is fixed, or given, the entropy of E (corresponding to that of the outcome) is not to be less than that of D, i.e.

$$H_R(E) \geqslant H_R(D)$$

Now whatever the causal or other relations between D, R, and E, algebraic necessity requires that their entropies must be related so that:

$$H(D) + H_D(R) = H(R) + H_R(D)$$

for each side of the equation equals $H(R, D)$. Substitute $H_R(E)$ for $H_R(D)$, and we get

$$\begin{aligned} H(D) + H_D(R) &\leqslant H(R) + H_R(E) \\ &\leqslant H(R, E) \end{aligned}$$

But always, by algebraic necessity,

$$H(R, E) \leqslant H(R) + H(E)$$

so

$$H(D) + H_D(R) \leqslant H(R) + H(E)$$

i.e.

$$H(E) \geqslant H(D) + H_D(R) - H(R).$$

Thus the entropy of the E's has a certain minimum. If this minimum is to be affected by a relation between the D- and R-sources, it can be made least when $H_D(R) = 0$, i.e. *when R is a determinate function of D*. When this is so, then $H(E)$'s minimum is $H(D) - H(R)$, a deduction similar to that of the previous section. It says simply that the minimal value of E's entropy can be forced down below that of D only by an equal *increase* in that of R.

1.9. The theorems just established can easily be modified to give a worth-while extension.

Consider the case when, even when R does nothing (i.e. produces the same move whatever D does) the variety of outcome is *less* than that of D. This is the case in Table 2. Thus if R gives the reply α to all D's moves, then the outcomes are a, b, or d – a variety of three, less than D's variety of five. To get a manageable calculation, suppose that within each column each element is now repeated k times (instead of the 'once only' of 1.5). The same argument as before, modified in that kn rows may provide only one outcome, leads to the theorem that

$$V_O \geqslant V_D - \log k - V_R$$

in which the varieties are measured logarithmically.

An exactly similar modification may be made to the theorem in terms of entropies, by supposing, not as in 1.8 that

$$H_R(E) \geqslant H_R(D), \text{ but that}$$
$$H_R(E) \geqslant H_R(D) - K$$

$H(E)$'s minimum then becomes

$$H(D) - K - H(R)$$

with a similar interpretation.

1.10. The law states that certain events are impossible. It is important that we should be clear as to the origin of the impossibility. Thus, what has the statement to fear from experiment?

It has nothing to do with the properties of matter. So if the law is stated in the form, 'No machine can . . .', it is not to be overthrown by the invention of some new device or some new electronic circuit, or the discovery of some new element. It does not even have anything to do with the properties of the machine in the general sense of Chapter 4 [not included here]; for it comes from the *table*, such as that of 1.4; this table says simply that certain *D–R* combinations lead to certain outcomes, but is quite independent of whatever it is that determines the outcome. Experiments can only *provide* such tables.

The theorem is primarily a statement about possible arrangements in a rectangular table. It says that certain types of arrangement cannot be made. It is thus no more dependent on special properties of machines than is, say, the 'theorem' that four objects can be arranged to form a square while three can not. The law therefore owes nothing to experiment.

1.11. *Regulation again.* We can now take up again the subject of regulation, ignored since the beginning of this chapter, for the Law of Requisite Variety enables us to apply a *measure* to regulation. Let us go back and reconsider what is meant, essentially, by 'regulation'.

There is first a set of disturbances *D*, that start in the world outside the organism, often far from it, and that threaten, if the regulator *R* does nothing, to drive the essential variables *E* outside their proper range of values. The values of *E* correspond to the 'outcomes' of the previous sections. Of all these *E*-values only a few (η) are compatible with the organism's life, or are unobjectionable, so that the regulator *R*, to be successful, must take its value in a way so related to that of *D* that the outcome is, if possible, always within the acceptable set η, i.e. within physiological limits. Regulation is thus related fundamentally to the game of 1.4. Let us trace the relation in more detail.

The Table *T* is first assumed to be given. It is the hard external world, or those internal matters that the would-be regulator has to take for granted. Now starts a process. *D* takes an arbitrary value, *R* takes some value determined by *D*'s value, the table determines an outcome, and this either is or is not in η. Usually the process is repeated, as when a thermostatically heated water-bath deals, during the day, with various disturbances. Then another value is taken by *D*, another by *R*, another outcome occurs, and this also may be either in η or not. And so on. If *R* is a well-made regulator – one that works successfully – then *R* is such a transformation of *D* that all the outcomes fall within η. *In this case R and T together are acting as the barrier F* [paragraph 10.5 – not included here].

We can now show these relations by the diagram of immediate effects (Figure 1).

The arrows represent actual channels of communication. For the variety in *D* determines the variety in *R*; and that in *T* is determined by that in both *D* and *R*. If *R* and *T* are in fact actual machines, then *R* has an input from *D*, and *T* has two inputs.

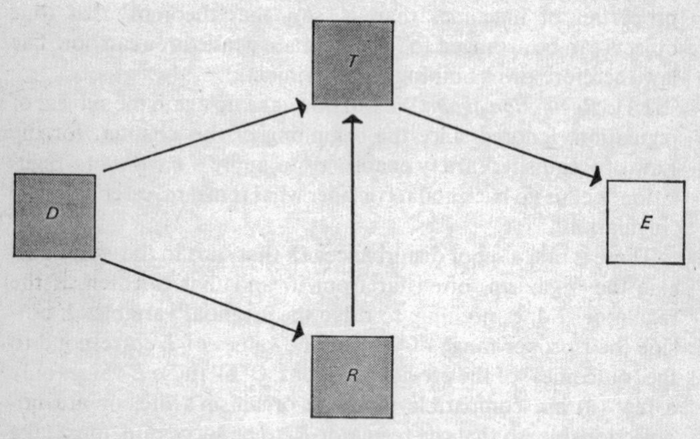

Figure 1

(When *R* and *T* are embodied in actual machines, care must be taken that we are clear about what we are referring to. If some machine is providing the basis for *T*, it will have a set of states that occur step by step. These states, and these steps, are essentially independent of the discrete steps that we have considered to be taken by *D*, *R*, and *T* in this chapter. Thus, *T* gives the outcome, and any particular outcome may be compared with another, as unit with unit. Each individual outcome may, however, in another context, be analysed more finely. Thus a thirsty organism may follow trajectory 1 and get relief, or trajectory 2 and die of thirst. For some purposes the two outcomes can be treated as units, particularly if they are to be contrasted. If however we want to investigate the behavior in more detail, we can regard trajectory 1 as composed of a sequence of states, separated by steps in time that are of quite a different order of size from those between successive regulatory acts to successive disturbances.)

We can now interpret the general phenomenon of regulation in terms of communucation. If *R* does nothing, i.e. keeps to one value, then the variety in *D* threatens to go through *T* to *E*, contrary to what is wanted. It may happen that *T*, without change by *R*, will block some of the variety (paragraph 1.9), and occasionally this blocking may give sufficient constancy at *E* for survival. More commonly, a further suppression at *E* is necessary; it can be achieved, as we saw in paragraph 1.6 only by further variety at *R*.

We can now select a portion of the diagram, and focus attention on *R* as a transmitter (Figure 2).

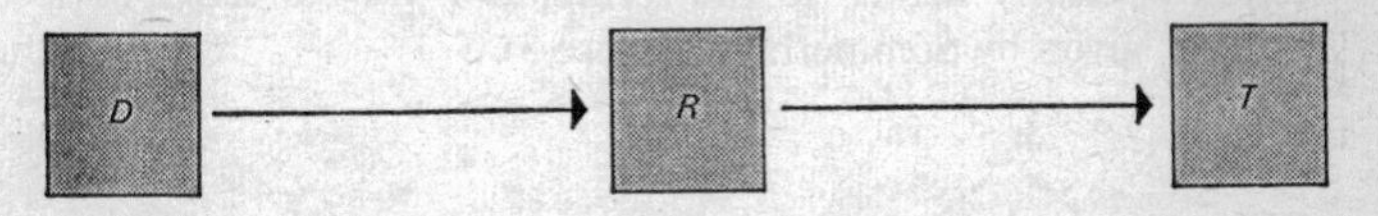

Figure 2

The Law of Requisite Variety says that *R's capacity as a regulator cannot exceed R's capacity as a channel of communication.*

In the form just given, the Law of Requisite Variety can be shown in exact relation to Shannon's Theorem 10, which says that if noise appears in a message, the amount of noise that can be removed by a correction channel is limited to the amount of information that can be carried by that channel.

Thus, his 'noise' corresponds to our 'disturbance', his 'correction channel' to our 'regulator *R*', and his 'message of entropy *H*' becomes, in our case, a message of entropy zero, for it is *constancy* that is to be 'transmitted'. Thus the use of a regulator to achieve homeostasis and the use of a correction channel to suppress noise are homologous.

Exercise 1. A certain insect has an optic nerve of a hundred fibers, each of which can carry twenty bits per second; is this sufficient to enable it to defend itself against ten distinct dangers, each of which may, or may not, independently, be present in each second?

Exercise 2. A ship's telegraph from bridge to engine-room can determine one of nine speeds not oftener than one signal in five seconds, and the wheel can determine one of fifty rudder positions in each second. Since experience has shown that this means of control is normally sufficient for full regulation, estimate a normal upper limit for the disturbances (gusts, traffic, shoals, etc.) that threaten the ship's safety.

Exercise 3. A general is opposed by any army of ten divisions, each of which may manoeuvre with a variety of 10^6 bits in each day. His intelligence comes through ten signallers, each of whom can transmit sixty letters per minute for eight hours in each day, in a code that transmits two bits per letter. Is his intelligence channel sufficient for him to be able to achieve complete regulation?

Exercise 4. (Continued.) The general can dictate orders at 500 bits/minute for twelve hours/day. If his Intelligence were complete, would this verbal channel be sufficient for complete regulation?

1.12. The diagram of immediate effects given in the previous section is clearly related to the formulation for 'directive correlation' given by Sommerhoff (Figure 3).

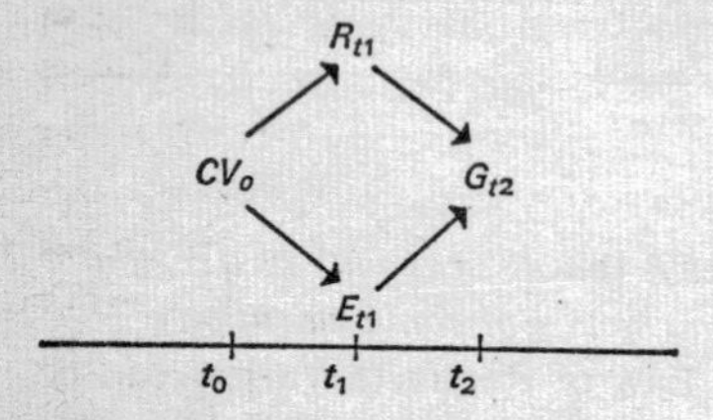

Figure 3 (From G. Sommerhoff, *Analytical Biology*, Oxford University Press, 1950)

If I am not misinterpreting Sommerhoff, his concepts and those used here are equivalent thus:

Coenetic variable (CV_O) ↔ Disturbance (D)
Response (R_{t1}) ↔ Response (R)
Environmental circumstances (E_{t1}) ↔ Table (T)
Subsequent occurrence (G_{t2}) ↔ Outcome (E)

A reading of his book [see caption to Figure 3] may thus help to extend much of the theory given in this Part, for he discusses the subject extensively.

1.13. The law now enables us to see the relations existing between the various types of variety and information that affect the living organism.

A species continues to exist (paragraph 10.4) primarily because its members can block the flow of variety (thought of as disturbance) to the gene pattern (paragraph 10.6), and this blockage is the species' most fundamental need. Natural selection has shown the advantage to be gained by taking a large amount of variety (as information) partly into the system (so that it does not reach the gene pattern) and then using this information so that the flow via *R* blocks the flow through the environment *T*.

This point of view enables us to resolve what might at first seem a paradox – that the higher organisms have sensitive skins, responsive nervous systems, and often an instinct that impels them, in play or curiosity, to bring more variety to the system than is immediately necessary. Would not their chance of survival be improved by an avoidance of this variety?

The discussion in this chapter has shown that variety (whether information or disturbance) comes to the organism in two forms. There is that which threatens the survival of the gene pattern – the direct transmission by *T* from *D* to *E*. This part must be blocked at all costs. And there is that which, while it may threaten the gene pattern, can be transformed (or re-coded) through the regulator *R* and used to block the effect of the remainder (in *T*). This information is useful, and should (if the regulator can be provided) be made as large as possible; for, by the Law of Requisite Variety, the amount of disturbance that reaches the gene pattern can be diminished only by the amount of information so transmitted. That is the importance of the law in biology.

It is also of importance to *us* as we make our way towards the last chapter. In its elementary forms the law is intuitively obvious and hardly deserving statement. If, for instance, a press photographer would deal with twenty subjects that are (for exposure and distance) distinct, then his camera must obviously be capable of at least twenty distinct settings if all the negatives are to be brought to a uniform density and sharpness. Where the law, in its quantitative form, develops its power is when we come to consider the system in which these matters are not so obvious, and particularly when it is very large. Thus, by how much can a dictator control a country? It is commonly said that Hitler's control over Germany was total. So far as his power of regulation [in the sense of paragraph 10.6 – not included here] was concerned, the law says that his control amounted to just one man-power, and no more. (Whether this statement is true must be tested by the future; its chief virtue now is that it is exact and uncompromising.) Thus the law, though trite in the simple cases, can give real guidance in those cases that are much too complex to be handled by unaided intuition.

Control

1.14. The formulations given in this chapter have already suggested that regulation and control are intimately related. Thus, Table 1 enables R not only to achieve a as outcome in spite of all D's variations; but equally to achieve b or c at will.

We can look at the situation in another way. Suppose the decision of what outcome is to be the target is made by some controller, C, whom R must obey. C's decision will affect R's choice of α, β, or γ; so the diagram of immediate effects is as in Figure 4.

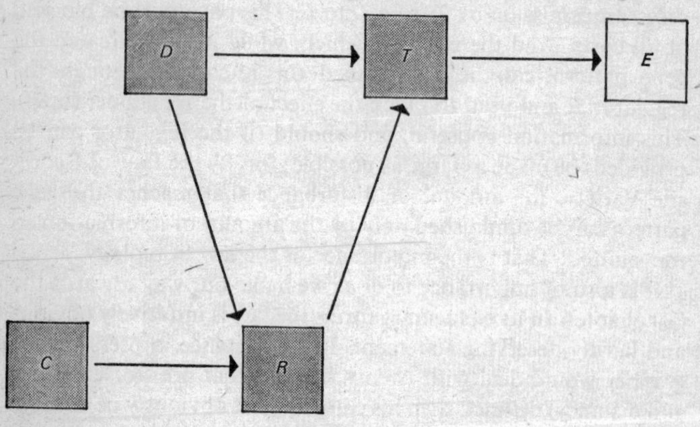

Figure 4

Thus the whole represents a system with two independent inputs, C and D.

Suppose now that R is a perfect regulator. If C sets a as the target, then (through R's agency) E will take the value a, *whatever value D may take*. Similarly, if C sets b as target, b will appear as outcome whatever value D may take. And so on. And if C sets a particular sequence – *a*, *b*, *a*, *c*, *c*, *a*, say – as sequential or compound target, then that sequence will be produced, regardless of *D*'s values during the sequence. (It is assumed for convenience that the components move in step.) Thus the fact that *R* is a perfect regulator gives *C* complete control over the output, in spite of the entrance of disturbing effects by way of *D*. Thus, *perfect* regulation *of the outcome by R makes possible a complete* control *over the outcome by C.*

We can see the same facts from yet another point of view. If an attempt at control, by *C* over *E* (Figure 5),

Figure 5

is disturbed or made noisy by another, independent, input *D*, so that the connexions are as in Figure 6,

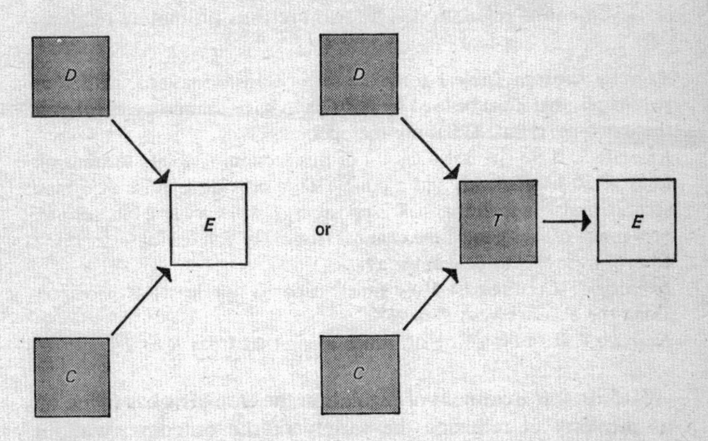

Figure 6

then a suitable regulator *R*, taking information from both *C* and *D*, and interposed between *C* and *T*:

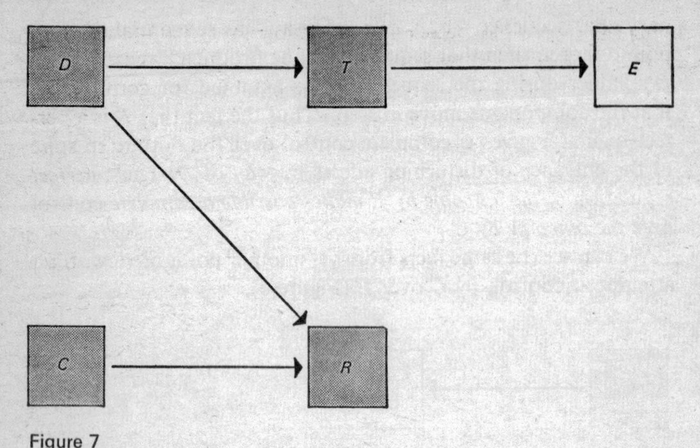

Figure 7

may be able to form, with *T*, a compound channel to *E* that *transmits fully from C while transmitting nothing from D.*

The achievement of control may thus depend necessarily on the achievement of regulation. The two are thus intimately related.

Exercise 1. From Table 1 form the set of transformations, with *C* as parameter, that must be used by *R* if *C* is to have complete control over the outcome. (Hint: What are the operands?)
Exercise 2. If, in the last diagram of this section, *C* wants to transmit to *E* at 20 bits/second, and a source *D* is providing noise at 5 bits/second, and *T* is such that if *R* is constant, *E* will vary at 2 bits/second, how much capacity must the channel from *D* to *R* have (at least) if *C*'s control over *E* is to be complete?
Exercise 3. (Continued.) How much capacity (at least) is necessary along the channel from *C* to *R*?
Exercise 4. (Continued.) How much along that from *R* to *T*?

1.15. In our treatment of regulation the emphasis has fallen on its property of reducing the variety in the outcome; without regulation the variety is large – with regulation it is small. The limit of this reduction is the regulation that holds the outcome rigorously constant. This point of view is undoubtedly valid, but at first it may seem to contrast sharply with the naïve view that living organisms are, in general, anything but immobile. A few words, in addition to what was said in 1.13, may be useful.

It should be appreciated that the distinction between 'constant' and 'varying' often depends on the exact definition of what is being referred to. Thus if a searchlight follows an aircraft accurately we may notice either that the searchlight moved through a great range of angles (angles in relation to the earth) or that the angle it made with the aircraft remained constant at zero. Obviously both points of view are valid; there is no real contradiction in this example between 'great range' and 'constant', for they refer to different variables.

Again, the driver who steers a car accurately from one town to another along a winding lane can be regarded either as one who has caused the steering wheel to show much activity and change or as one who, throughout the trip, has kept the distance between car and verge almost constant.

Many of the activities of living organisms permit this double aspect. On the one hand the observer can notice the great deal of actual movement and change that occurs, and on the other hand he can observe that throughout these activities, so far as they are coordinated or homeostatic, there are invariants and constancies that show the degree of regulation that is being achieved.

Many variations are possible on the same theme. Thus if variable x is always doing just the same as variable y, then the quantity $x - y$ is constant at zero. So if y's values are given by some outside factor, any regulator that acts on x so as to keep $x - y$ constant at zero is in fact forcing x to vary, copying y. Similarly, 'making x do the opposite to y' corresponds to 'keeping $x + y$ at some constant value'. And 'make the variable w change so that it is always just twice as large as v's (fluctuating) rate of change' corresponds to 'keep the quantity $w - 2dv/dt$ constant'.

It is a great convenience in exposition and in the processes of general theory to be able to treat all 'targets' as if they were of the form 'keep the outcome constant at a'. The reader must, however, not be misled into thinking that the theory treats only of immobility; he must accustom himself to interchanging the corresponding concepts freely.

Some Variations

1.16. In 1.4 the essential facts implied by regulation were shown as a simple rectangular table, as if it were a game between two players *D* and *R*. The reader may feel that this formulation is much too simple and that there are well-known regulations that it is insufficient to represent. The formulation, however, is really much more general than it seems, and in the remaining sections of this chapter we shall examine various complications that prove, on closer examination, to be really included in the basic formulation of 1.4.

1.17. *Compound disturbance*. The basic formulation of 1.4 included only one source of disturbance *D*, and thus seems, at first sight, not to include all those cases, innumerable in the biological world, in which the regulation has to be conducted against several disturbances coming simultaneously by several channels. Thus, a cyclist often has to deal both with obstructions due to traffic and with disequilibrations due to gusts.

In fact, however, this case is included; for nothing in this chapter excludes the possibility that *D* may be a vector, with any number of components. A vectorial *D* is thus able to represent all such compound disturbances within the basic formulation.

1.18. *Noise*. A related case occurs when *T* is 'noisy' – when *T* has an extra input that is affected by some disturbance that interferes with it. This might be the case if *T* were an electrical machine, somewhat disturbed by variations in the mains voltage. At first sight this case seems to be not represented in the basic formulation.

It must be appreciated that *D*, *T*, *E*, etc., were defined in 1.3 in purely *functional* form. Thus '*D*' is 'that which disturbs'. Given any real system some care may be necessary in deciding what corresponds to *D*, what to *T*, and so on. Further, a boundary drawn provisionally between *D* and *T* (and the other boundaries) may, on second thoughts, require moving. Thus one set of boundaries on the real system may give a system that purports to be of *D*, *T*, etc., yet does not agree with the basic formulation of 1.4. Then it may be found that a shifting of the boundaries, to give a *new D*, *T*, etc., gives a set that *does* agree with the formulation.

If a preliminary placing of the boundaries shows that this (provisional) T is noisy, then the boundaries should be re-drawn so as to get T's input of noise [paragraph 9.19 – not included here] included *as a component in D*. D is now 'that which disturbs', and T has no third input; so the formulation agrees with that of 1.4.

There is, of course, no suggestion here that the noise, as a disturbance, can be allowed for magically by merely thinking differently about it. The suggestion is that if we start again from the beginning, and re-define D and T then some *new* transformation of D may be able to restore regulation. The new transformation will, of course, have to be more complex than the old, for D will have more components.

1.19. *Initial states.* A related case occurs when T is some machine that shows its behavior by a trajectory, with the outcome E depending on the properties of T's trajectory. The outcomes will then usually be affected by which of T's states is the initial one. How does T's initial state come into the basic formulation of 1.4?

If the initial state can be controlled, so that the trajectory can be started always from some standardized state, then no difficulty arises. [In this connexion the method of paragraph 7.25, not included here, may be useful.] It may however happen, especially if the system is very large, that T's initial state cannot be standardized. Does the basic formulation include this case?

It does; for D, as a vector, can be re-defined to include T's initial state. Then the variety brought to E by the variety in T's initial state is allotted its proper place in the formulation.

1.20. *Compound target.* It may happen that the acceptable states η at E may have more than one condition. Thus of a thermostat it might be demanded that:

(i) it shall usually stay between 36° and 37°C;
(ii) if displaced by $\pm 10°$ it shall return to the allowed range within one minute.

This difficulty can be dealt with by the same method as in 1.17, by recognizing that E may be a vector, with more than one component, and that what is acceptable (η) may be given in the form of separate specifications for each component.

Thus by allowing E to become a vector, the basic formulation of 1.4 can be made to include all cases in which the target is complex, or conditional, or qualified.

1.21. *Internal complexities.* As a last example, showing how comprehensive the basic formulation really is, consider the case in which the major problem seems to be not so much a regulation as an interaction between several regulations. Thus a signalman may have to handle several trains coming to his section simultaneously. To handle any one by itself would be straightforward, but here the problem is the control of them as a complex whole pattern.

This case is in fact still covered by the basic formulation. For nothing in that formulation prevents the quantities or states or elements in D, R, T, or E from being made of parts, and the parts interrelated. The fact that 'D' is a single letter in no way implies that what it represents must be internally simple or unitary.

The signalman's 'disturbance' D is the particular set of trains arriving in some particular pattern over space and time. Other arrangements would provide other values for D, which must, of course, be a vector. The outcomes E will be various complex patterns of trains moving in relation to one another and moving away from his section. The acceptable set η will certainly include a component 'no collision' and will probably include others as well. His responses R will include a variety of patterns of movements of signals and points. T is what is given – the basic matters of geography, mechanics, signalling techniques, etc., that lead determinately from the situation that has arisen and his reaction pattern to outcome.

It will be seen therefore that the basic formulation is capable, in principle, of including cases of any degree of internal complexity.

7 V. I. Kremyanskiy

Certain Peculiarities of Organisms as a 'System' from the Point of View of Physics, Cybernetics, and Biology

V. I. Kremyanskiy, 'Certain peculiarities of organisms as a "system" from the point of view of physics, cybernetics, and biology', *General Systems*, vol. 5 (1960), Society for General Systems Research, pp. 221–30. [This paper first appeared in Russian in *Voprosy Filosofii*, August (1958), pp. 97–107.]

In the heated discussion going on in our country about the significance of similarities and differences between automatic electronic devices and living organisms, the cyberneticists have usually been reproached for their extreme universalism, and also, on occasion, for their inadequate knowledge of biology. The opposition has, on the other hand, often pointed to the indistinctiveness of organisms or to such distinctive qualities of theirs which have already or could become properties of cybernetic machines.

The present article represents a biologist's attempt to introduce greater accuracy into the posing of certain questions and into the defining of the distinctions of living entities regarded as material systems of a special type.

Which Organisms are Involved?

In the discussion of these problems most of the attention has been given to those plans of automatic machines of modulating systems which are designed to increase the effectiveness of the brain or to study it. As a rule, reference is made only to the nervous system of humans and the higher animals. It is not difficult to see that this limitation is broadened at the discussion in question. The reasons for it are clear – these devices are, after all, designed to raise the productivity of mental labor.

Nevertheless, the question of the significance of the similarities is often posed in a most general form. For example, can cybernetic automatic devices be acknowledged as 'living'? Biologists are often disinclined to admit these machines into their domain. Is it possible, however, to solve the problem of understanding organic life at all if as evidence we use facts relating to society, i.e.

another form of material existence, or only to the highest forms of activity of the highest animals? In that case the concept 'living entity' does not include the lower animals and those organisms which have no nervous system – the plants and most of the microbes.

Within the organic world itself, there evidently exist very profound qualitative differences between different organisms. It is a universally known fact that life is possible only in highly organized bodies; but natural scientists cannot, of course, be satisfied with general ideas concerning the higher or lower organization of living bodies. There exists a vast literature dealing with matters concerning the basic levels or degrees of organization which have developed in the organic world and beyond its limits, the peculiarities characteristic of each of these degrees, and the reason that one level is indeed 'higher' or 'lower' than another. However, there does not yet exist a strictly objective criterion of progress. Without dwelling upon the history of these problems or indulging in unfavorable criticism of the respective theories and nomenclatures, we attempt primarily to give a very brief definition of the most important differences between the fundamental types of material systems.

The Fundamental Types of Material Systems

In mathematical logic there is a theory of numbers which concerns the grouping of objects, phenomena, and concepts, including those whose combination might be hypothetical or abstract. Contrary to this theory, the theories of material systems pertain solely to actually existing associations. One of the simplest types of association includes such things, for example, as undifferentiated gas and dust mist or small amounts of gaseous substances at 'moderate' temperature and pressure, which are not in eddy or stream movement. Associations of this type might properly be called unorganized systems or chaotic aggregates.

It would be wrong to think that such systems contain no unity or integrity, or that the properties of a chaotic aggregate are no different from the simple sum or simple physical resultant of the properties of its elements. Contrary to the view widely held among foreign scientists, no material association is absolutely additive in this sense. The existence of any relationship between two objects or two changes creates a situation which could not exist in either of them separately. It is true, however, that from a number of important aspects, or from a single aspect, the properties of the chaotic aggregate may not actually differ from the simple sum of the components' properties.

In chaotic aggregates, the interconnexions between the elements are comparatively uniform, but they are particularly simple when a relative organizational simplicity typifies the elements, themselves. The nature of the elements is not changed by entering or leaving the aggregate. Where there is a large number of elements, changes in the chaotic whole depend more on changes in many elements than on solitary or small groups of elements. The total of its internal interconnexions, and hence the internal causal conditioning of the changes, bears a predominantly statistical or 'probability' character. Not only in the inorganic world, but also in the organic world (and even in society), do relationships of this type exist. But it is only in chaotic aggregates that they are the leading relationships and the main basis for the most fundamental changes in the aggregate. (The idea of a main basis or leading moment should not be confused with the idea of the historically primary basis, to which we will come back later.)

Numbers can belong to classes of different degree. The class of each successive degree includes a certain number of classes from the preceding degree as their members. This is also true of material systems. It should be noted that when studying material associations in the various degrees one need not carry the analysis down to 'last elements' (whose existence we will leave aside), but only to a given number of preceding degrees of complexity.

Thus number of degrees is not always the same. When studying chaotic aggregates it is usually sufficient to carry the analysis to the elements representing the objects and changes of one preceding degree. Thus, when studying the regularity of changes in a small amount of unsuperheated gas, it is usually sufficient to carry the analysis to the molecules, viewing them as 'simple' particles. It is a different matter when studying material associations of other types. The more developed their internal and external connexions, and the more complex the partial systems (subsystems) forming the material association, the more the whole is dependent upon the individual components; or, in other words, the deeper the analysis must penetrate in order to reveal the details which are needed to understand the whole.

Having taken cognizance of these incomplete definitions of the peculiarities of chaotic aggregates, let us proceed further. In contrast to chaotic aggregates, organized systems are typified by basically regulated, varied and deep-seated internal connexions among the elements. Since the latter now exist in different relationships and can alter fundamentally, it would be better to find another term for 'elements'. In the 'general theory of systems', which is now popular in the United States of America through the initiative of L. von Bertalanffy, the word 'systems' is used only for organized associations, the members of the systems being called 'objects'. But for a number of reasons we think it preferable to use the term components, an expression which has long been used in describing physical systems and automatic devices.

The more varied and complex the interconnexions between components or subsystems (groups of components), the deeper the changes in the components (usually in only the first or second immediately preceding degrees). But these components can change only to the extent of their inherent capacity for change. For example, atoms change in molecules, and inorganic molecules change in crystals, solutions, and cells; but there is far greater change in large polymerized molecules (macromolecules) in cells, and cells in multicellular organisms. The most profound changes occur in multicellular animals in the higher-degree systems.

Furthermore, the essential features of the components can do more than change. They can be newly created through the creative capacities of the system (or subsystem). For example, the overwhelming majority of complex organic substances are synthesized only in the cell organelles, i.e. in the cell 'subsystems', under the influence of enzymes which are disposed and which act in a definite order. Analogous examples are numerous in nonliving nature as well, but here there are important differences which are usually either totally ignored or are given inadequate attention by the authors of existing systems theories.

Very serious attention is given in many branches of natural

science, in cybernetics and in the general theory of systems, to the relationship between the system and the environment. Below, we will trace out several very fundamental differences in this relationship, which radically change its nature at the various levels of development, but for the moment we will point up only the most general aspects. The environment includes the objects and changes which exert considerable influence on the material system without being part of it. Account must also be taken of influences which become considerable in their totality rather than separately as, for example, the effects of extremely distant cosmic bodies. The environment furthermore includes all objects and phenomena which feel the strong and direct (or not too remotely mediated) effects of the system.

Material systems are called, according to their type of relationship with the environment, isolated, closed or open. Absolutely isolated systems are, of course, purely abstract and hypothetical. In the changes of closed systems, the exchange of elements and energy with the environment does not play a very important role for a very long time. The entropy of the closed system as a rule only grows, whereas the system as a whole, being subservient to the environment and incapable of renewing itself, is inevitably destroyed, without, moreover, leaving a successor. After systems of this type, no matter how sturdy their structure, degenerate and disintegrate, other systems of this type must emerge as systems, anew. However, the succession of 'embryos' can also have considerable significance in nonliving matter, such as the 'primer' in crystallization.

In open systems, on the other hand, a periodic or continuous exchange of elements and energy with the environment is typical. This exchange can serve as the basis for the perpetuation of this form of existence and as the basis for the decrease or relative constancy of entropy only when the system possesses certain features of internal organization and interaction with the environment. There is no need to explain that these are then living bodies, and that they contain far more regulated internal interconnexions which associate the components so intimately that the system becomes, as is said, an 'organic whole'. This does not necessarily signify that there are organic substances containing carbon or silicon in its make-up. To avoid confusion it would be better to call material systems of this type differently, e.g. organic systems, with this notion also including those open systems which cannot definitely be classified either as living or as consisting of organic substances.

The Relationships Between Different Degrees of Organizational Development of Systems

Following the rather widespread custom, we will call these degrees the organizational orders of matter. The relationship between the organizational orders of matter consists in that the organic system of each succeeding order contains the systems of the preceding order as its basic components, not directly, however, but mainly as part of the subsystem, e.g. as part of the organelles of cells, or organs of multicellular organisms.

Here the whole contains the parts 'under itself', as Hegel said. The whole is bigger than the sum but not bigger than the organized system of its parts, in all their connexions and intermediaries. In the organic world these latter are extremely complex. We know that in all complex combinations of interconnexions we must be able to isolate the leading link. But to do this we must make a careful study of all the details having essential significance for the entire system. Through how many degrees should biological analysis be carried? We can immediately say through more than one, since this would be sufficient only in the case of chaotic aggregates or only in the narrow examination, e.g. when studying the birth rate in certain rodents, insects, etc. But these are not the major preoccupations of biology. When studying the most important phenomena of organic life it is mandatory to delve into the 'smaller-scale' phenomena whose significance increases as the level of material organization heightens. Hence, such sciences as cytology, biochemistry and biophysics are not only vitally pertinent in studies of microbes, but also in studies of multicellular organisms. It is not enough simply to 'recognize' these sciences; their achievements and methods must be put to as wide use as possible. Indeed, the under-estimation of details has caused our cytological and biophysical research to fall behind perceptibly, a situation which should be rectified as quickly as possible.

Those subsystems (organs and organ systems) which form the

leading link in the particular living system as a rule also serve as the main link for its interconnexions with the environment. This is due, of course, to the fact that plants and animals dominate only over secondary spheres in their environment (which are ever-expanding, however) and are obliged for the most part to adapt themselves to the environment. In special conditions of embryonic development this rule is broken, and the leading role can temporarily be played by the subsystems which in these stages do not perform the functions of interconnexion with the environment; but this is only possible in so far as this embryonic development is ensured by the previous generation's struggle with the environment, by means of its accumulated energy and materials.

Not only the properties and changes of the organic system as a whole, but also the specific properties of its major subsystems, are completely absent in the independently existing bodies of the type of the system's individual components. It is a different matter when bodies of this type unite and develop into a system of the following order. In the course of development, as we mentioned above, the nature of the components may take on the basic 'imprint' of the whole. Thus, to compare them with systems of subsequent orders, we must not select their components, but the separately existing bodies of the type of these components. When comparing organic systems of the same order, the highest significance may be had by the distinctions acquired by these systems as a result of their being included or newly formed in systems of higher orders.

It follows from this, incidentally, that the adherence of an animal and automatic device to the same type of organic system would not signify a truly fundamental similarity between them. A computing machine can possess many of the basic features of an animal, but the former represents a component of society which was created newly by man (or by other automatic devices, the possibility of which was pointed out by von Neumann).

The developed components of an organic whole always contain something more than they would alone, but never more than the given whole and its environment (and also its history). This is not taken into consideration by those foreign cyberneticists, and particularly philosophers, who arrive at idealist conclusions by supposing that the information involved in the operation of, say, memory devices or communication channels is something that can be reduced neither to matter nor energy. The idea that this information is hence something 'non-material' is very much like the 'overcoat point of view', in which only the trappings are considered matter, while society with its internal and external relationships is not considered matter. The word matter should not be reduced to the word substance in its physicochemical sense, as is generally the trend outside the sphere of gnoseology.

Let us now see how the problems raised at the beginning of this article could be handled on the basis of the aforesaid. Let us first frame an objective criterion for progress, and then outline the relative importance of the similarities and differences between living systems of the organizational orders, representing the most important stages of development in the organic world.

But we first must prove in a more complete exposition that the leading properties of organic systems are actually superior in certain stages of development to the subordinate properties or subordinate variables (where domination does not always mean superiority) and that the development of ways and means of variability offers the organism more, and not fewer, advantages than the development of dynamic stability. We note that cyberneticists emphasize in theory the significance of ultrastability and multistability (phenomena discovered by Ashby), but in practice they also concern themselves with variability. It is known that the latter serves as a means for attaining stability (stabilizing variability); but, in addition (and this is far more important for biological systems), it serves as a means for developing new forms both of the species' stability and also of the variability itself (form-creating variability). It is easily shown that an increased number of components (the limit to which is specific for each order depending upon the available integration means and upon the environment conditions) increases variability and has several other advantages, of which we will speak in time. There apparently exists a phenomenon of 'ultravariability' (the repeated and qualitatively heightened capacity for change).

In the light of these preliminary schematic observations and that which was said earlier, we can objectively, if in 'qualitative' form only, draw the following conclusions with regard to the questions posed:

1. One organic system is truly superior to another organic system if (a) the first possesses every essential property of the second, but in addition to this relationship of essential similarity or kinship (not to be confused with the relationship of identity) with the latter, it possesses properties which are essentially lacking in the second, whereupon (b) the total of these original properties of the first, with the pre-eminence of one of them, acts as the leading factor in its internal and external interconnexions, that is, relates to the properties of the second system as major to secondary (and, in the end, better ensures preservation and development).

It is easily seen that it is just this relationship that does indeed exist between the organic systems of each succeeding order and the systems of each preceding order. Consequently, the organic system of each succeeding order is indeed superior to the system of each preceding order (when certain conditions exist, which should also be defined in a more complete way).

Organic life has passed through at least three major stages of development in types of transition to higher orders of organizations (cells, multicellular organisms, and family-herd groups). Since the relationship of organizational orders is the same as the relationship of progress, it becomes impossible to deny the progressive nature of development in the organic world. (There are still biologists and philosophers who deny the existence of progressive development.) All progress is relative, and bound up with regression (Engels), but progress can relate to the leading properties or major moments in a form of existence, while regression can relate to the secondary moments and subordinate properties. Then, with all the relativity of progress, the development as a whole proves to be progressive. True progress enriches the forms of existence.

We are of the opinion that the elaboration of a system of concepts connected with this or some other objective criterion of superiority, might open the way to calculating progress, i.e. to a technique for the strictly objective and quantitative determination of the degree of true progressiveness in new or planned changes.

2. When comparing organic systems of different orders, the exclusion of everything similar results in the isolation of everything which does not belong to the systems of the lower order.

These distinctions are precisely what we must know about the living systems of the higher order, for it is in these distinctions that the leading moments of their life-forms lie. Thus, if we are comparing organic systems of different orders of organization, the differences between them are of greater importance than the similarities.

Corresponding to these profound differences between the most important degrees of progressive development of organic life are the highly essential differences in the types of relationship of living bodies to their own existence.

Forms of 'Feedback' in Living Nature

Self-awareness (in a certain sense the 'reflexive relationship' in the theory of numbers corresponds to it) in living systems is based in particular, on the progress development of 'feedback'. Cybernetics has called attention to the part played by feedback in the organization of living bodies, after having created automatic devices modeling, in this regard as well, organisms of higher levels of development. Nor have exaggerations been lacking. Man has even been called 'a package of feedbacks' by some foreign cyberneticists (Ducroque).

Physiology studied feedback long before the appearance of cybernetics, and did not exaggerate its truly major importance. P. K. Anokhin, whose research has constituted a major contribution to the study of feedback in the nervous system, correctly noted in his article 'Physiology and cybernetics' (which appeared in *Problems of Philosophy*, no. 4, 1957) how restricted the point of view of many cyberneticists is. It needs to be supplemented primarily with the viewpoint of historical development, i.e. the arisal and evolution of feedback.

In the broad sense of the term, feedback is not only peculiar to organisms containing nervous systems. There exists, for example, a 'plus–minus interaction', studied in particular by the late M. M. Zavodovskiy (i.e. interconnexions between the functions of organs, in which the stronger functioning of one group of organs produces a more intensive functioning of those organs which exercise an inhibiting influence on activity of the former). This phenomenon can be produced otherwise than through a nervous system; it can also be produced through direct biochemical interactions between internal secretion organs joined by a 'cyclically' operating circulation system. Though relatively secondary in the developed organism of the higher animal, these and other biochemical or 'humoral' and morphophysiological (in embryonic development) types of feedback are in many cases of considerable practical significance, while in the early stages of onthogenesis and in the organisms of the lower animals (whose nervous systems have not yet begun to play the huge role in internal interconnexions that they do in the higher animals) biochemical and morphogenetic feedback is still the main form of internal feedback. In cells it is closer to the chemical forms of feedback, particularly to the Le Chatelier principle, according to which the accumulation of reverse reaction products changes the reaction's predominant direction to the opposite.

Feedback need not be only internal for individual living bodies. Not only as complements to internal feedback, but also above these latter do there exist means for evaluating phenomena and, more important, for evaluating their effects on initial, introductory changes in the organism's subsystems. These are the highly complicated processes showing the indirect influence not only of unequal abilities to survive, but also of different rates of development, multiplication and (or) of unequal abilities to use the elements and energy of the environment to react upon the environment and to supply the systems which perform these actions. These processes influence groups living in a single locality and united by a common succession of phylogenesis and also, more often than not, by a crossing of phylogenetic lines (populations, groups of populations, species). In a word, these are the processes of natural selection (a better translation of Darwin's idea than 'natural separation', which does not stress the idea of selection as much as the idea of simple sifting out).

Cybernetic machines, the deliberate creations of man, can be capable of self-reproduction. In this regard, they are quite similar to living things. Under present level of development of technology, it has been found better, apparently, to create these machines from the start in finished form. But even self-reproducing machines, remaining components of society, would not of course have to carry on a 'struggle for existence', as the rebellious robots in the famous play by Karel Capek. In a society which develops in planned rather than spontaneous fashion (as only communist society will be able to), only useful kinds of robots will be built and retained. To the extent that they are built freshly, an independent line of phylogenetic succession is not mandatory for them, as for natural living systems. But the question has already been raised of building machines with protracted self-reproduction, e.g. machines that might float at sea and extract substances useful to man from the water, while all the time multiplying and navigating back to factories where their bodies might be processed by other automatic devices.

The Evolution of the Relationship Between Organism and Environment

The question of the relationship between organism and environment, both internal and external, is one of the central problems in biology. There still exists today the trend of studying this problem on the basis of factual material relating to only a few forms of living bodies. The inadequate attention given to the necessity of having a differentiated and historical approach to this question is also typical of the viewpoint of the founders of cybernetics, just as for von Bertalanffy's 'general theory of systems'.

Actually here, too, is the simplest and lowest form of relations which is the most general (similar). They change radically at the various levels of progressive development. Let us examine several of the more important peculiarities of the relationship between a living system and environment at the most important levels of progressive development.

Regardless of how highly developed the self-awareness and independent activity of living systems, only the universe (nature and society, but not substance in the sense of building material or substratum) is an absolutely perfect 'cause unto itself'. Only the universe possesses complete self-motion. For all finite material formations and particularly for 'open' systems, there exists a relationship to the environment which is based on interconnexion with the environment, and hence there exists an interdependence between each system and its environment.

It has already been observed in our literature that unity and interconnexion in no way exclude opposition. There are periods, though, when direct unity of living systems with their environments does indeed exist. These are the periods of their arisal anew (without a successive tie with finished systems) from simpler elements which had not yet succeeded in becoming organic parts.

This was the case, for example, during the age of the first appearance of life on Earth. Organic life developed in solutions, the proper domain of chemistry. It is not yet clear what exact processes played the leading part in this development. Some of the suggestions have been: coacervation (Oparin), the formation of semi-liquid crystals (Lehman), the development of osmotic systems (it is not sufficiently clear how deep a similarity with certain forms of living bodies was possessed by the many osmotic forms obtained by Stefan Lediuk, some of which achieved metabolism and development), electrostatic and electrodynamic fields (Bastian, Krail, Lund), or several of these processes combined into one network (as appears to us most likely). But any of these processes would be sufficient to separate groups of complex and simple organic molecules which were previously dispersed in solution, together with part of the aqueous medium, from the rest of the solution. This did not occur within a single drop, but in an entire zone of primary conception, i.e. not in a single form.

When the primary living bodies had, in one way or other, come to life, they stood out as parts of solutions having more regulated internal interconnexions than those between the molecules and 'microcrystals' of water in the portions of the solutions where there were no such systems. In the relationship of the 'inside' and 'outside' that thus arose, the 'inside' of the primary living bodies began immediately to rise above the adjacent non-living 'outside', if only through their higher degree of complexity and organization (applying here the criterion of superiority given above).

Many primary living bodies, as A. I. Oparin emphasizes, died and arose anew; but there were also some which possessed the capacity for fairly protracted and continuous self-reproduction, and also of an entire chain of individual development cycles, or phylogenetic branchings, and later of a special way of aggregating these branchings – the species. It was precisely these that developed quickest. In the first processes of natural selection, there also arose and developed the many-sided internal successiveness of living bodies.

Thus, in even their very first stages of development, living bodies obtained a double advantage over their adjacent non-living environment – superiority of organization and superiority of succession (T. D. Lysenko). And as a result of their more highly developed successiveness and of their relative but real particularization of the 'inside', there began to accumulate selectively in the living bodies various results of their interaction with the outer environment. The selectivity of the metabolic processes, and assimilation itself, are obviously dependent more than anything else upon the 'inside' already accumulated. Having built up its 'inside' from the environment and under its influences, the living bodies, in each subsequent period of evolution, develop always in interconnexion with the environment, but no longer as completely passive objects of influence. They now emerge as increasingly active bodies and forces of nature, possessing their own potentialities for form-creation, as determined by the peculiarities of their increasingly complicated internal organization, which does not, however, preclude 'adventitious' and probability processes (the need for which in automatic device systems, in order to adapt themselves successfully, has been proven by cybernetics; this finding is of very great practical significance).

The non-living world remained as it was – a sphere of relatively disconnected and unorganized changes and movements of material formations. On the other hand, the interconnexions in living bodies came to possess better and better arranged chains of chemical and physicochemical reactions. Each fairly gentle external influence passing through the 'inside' of the living body immediately became a subordinate element with regard to the more highly organized system of processes. It is not difficult to defeat an enemy who marches into the camp of an entire army with only a small or weak detachment. And it is exactly this happy condition that is enjoyed by the most insignificant living body with regard to external conditions, if the outside influence (for example, some substance penetrating into it) does not destroy this relative superiority together with its possessor. Finding itself in new, but not radically different environmental conditions, the organism changes. It continues, however, to retain its selective activity with regard to the new external conditions, submitting to the influences of the environment only to the extent of its own capacity for change.

The metabolic processes and other interconnexions between organisms and environment became more and more complicated. Multicellular animals developed differentiated systems for seizing food, digestion, respiration, and internal transportation, the functioning of which depended more than ever on the connecting abilities of the ever-differentiating and concentrating nervous system. The directing influence of the nervous system extended to even the intracellular metabolic processes in almost all tissues of the organism (what we call the 'trophic' influences of the nervous system).

Furthermore, the nutrition of the higher multicellular animals with an 'active' way of life became not so much consumption, digestion and assimilations, not so much the activity of the 'internal environment' organs, as an external activity. The very functions of the cerebral cortex developed predominantly as means of ensuring and reflecting the processes of active adaptation to the environment.

In developed family-herd groups, the basic mediating link in the interconnexion with the environment, and hence the leading link in the internal interconnexions, is precisely the external and 'nervous' activity of the animal. The animals are separated from each other anatomically, but this is an illusory independence. One of the most striking zoological examples of the profound subordination of metabolism in multicellular organisms to the whole of the order is the ability of some ants to turn part of their group into living storehouses of honey. Mammals have not the developed anatomo-physiological differentiation in groups that colony insects have, but nevertheless the behavior of the individuals and the interconnexions in their behavior reach a higher zoological level.

Though external to individual animals, the outgoing actions of group members are at the same time internal for the group as a whole, i.e. they are links connecting its complex interconnexions. The relationship between outside and inside here is not only complex, but also many-staged. The external environment penetrates the entire living whole of such a group and turns in part into its internal environment, remaining external for the organisms of the group members.

In turn, the organisms of 'individuals' as multicellular beings possess their own internal environment which is external with regard to individual cells and some organs. But even cells are not monolithic in the physical and physiological sense; they also have their internal environment which is external for their basic components and subsystems (organelles). Hence, the idea of the 'relationship between inside and outside' should be differentiated depending on the organizational order of the given living whole.

As living systems of each such major degree arise, part of the external environment drawn into the developing association of living bodies which hitherto lived in a dispersed condition, becomes the internal environment of the new association as a whole. Thus, during the period of the beginning of the living systems of each subsequent order, the relationship between outside and inside once again, but newly, takes on the character of a direct and simple unity. But a truly fused unity of the living system and environment is intrinsic to this system *in statu nascendi* (in the nascent state) only when there is no successive connexion with a finished system of the same order. The processes of embryonic development occur in the higher animals in increasingly intimate physiological interconnexions with the mother organism, and these are subsequently complemented by successive interconnexions during nurturing and zoological rearing. All these complementary actions play the role of supercellular, 'supergenic' succession connexions between generations and also create essential differences in the relationships between organism and environment. The celebrated 'communication channels' which transmit 'genetic information' are not confined, in the higher animals, to the narrow framework of 'cellular bridges' between generations, as many physicists and cyberneticists think.

Cybernetic machine design is still in its earliest stages. But even in the simplest artificial homeostats (systems capable of maintaining dynamic stability despite 'disturbing' effects from without), there is not the direct, fused unity of system and environment, about which Ashby speaks, thus repeating in this respect the error of a number of biologists. There often is confusion between the need for various 'variables' in the environment without which the system cannot exist, and the leading role of the outside or inside. Conditionalism has revealed a tenacious durability in biology, and has even penetrated into cybernetics.

Evolutionists have long struggled with the question of the endogenous (where the leading factor is that of the independent reactive force of the organism, as Engels expressed it) or the octogenic (where the leading factor is that of changes in the environment) character of evolutionary processes. There is not yet sufficient clarity in this question. We would like to emphasize here the necessity of the historical and differentiated approach to the matter of the relationship between outside and inside, though clearly this is only 'a start in the matter'.

In going further into the question of the relationship between organism and environment, we would have to show also that the development of the relationship to the environment did not proceed along a straight line. The brain first developed as an organ of primarily internal interconnexion, and then primarily as an organ of interconnexions with the environment. This augmented the dependence of the organism on far more distant and complex events in the environment (with the development of the organs of perception and systems of receptors). The brain came to act as a Trojan horse by means of which the external environment penetrates into the very depths of the leading subsystem of the higher animal's organism, and from there, after having assumed the form of definite groups of its own internal impulses, directs its internal and external activity. Thus, it is profoundly mistaken to stress, as does J. Huxley for example, only the augmentation of the organism's 'independence' from the environment. This occurs in association with a deepening and widening mediated dependence. But in the highest development of means of adaptation to the environment there also lay the need for the opposite process – the development of new ways and means for combating the environment.

The domination of human society over its environment did not, of course, arise in 'empty space'. Questions involved in the 'transition from the leading role of natural selection to the leading role of labor' were taken up by us in another article ['Uspekhi sovermennoi biologii' (Advances of modern biology), 1941].

Properties of Open Systems

We must now follow the example of the physicists who are endeavoring to explain life and advance several ideas on certain physical aspects of biological systems, attempting here to describe primarily their peculiarities.

Life from the Standpoint of Physics

The physicists, who have in recent years taken up questions of biology, place great emphasis on the idea of the 'ectropism' of living bodies – the ability of open and organized systems to use the environmental substances, rich in easily released energy, to maintain irregular distributions (gradients) of substance concentrations and to support synthesis processes founded on the binding and simultaneously the dispersing of energy – in a word, to maintain a given level of entropy and even to lower it (this being expressed by the term 'negative entropy' or 'neg-entropy'). E. Shrödinger, in his book *What is Life?* (published by Cambridge University Press in 1944), attaches decisive significance to the existence in the cell and the participation in its metabolism of supermolecular solid structures, particularly the 'aperiodic crystals' of skeletal threads of chromosomes. These comparatively sturdy structures (physical chemists now talk more about the significance of macromolecule systems in protoplasm) are not subject, to the same extent as smaller molecules, to the dangers of random movements in solutions and hence, owing to their effects on biochemical reactions, can play the role of 'Maxwell's demons', creating order from disorder.

But osmotic membranes (whole systems of which were recently discovered in seemingly homogeneous cytoplasm by means of an electronic microscope), which act by a far more complex mechanism than was hitherto thought, can be no less worthy candidates for 'demons', while semi-liquid crystals or the boundaries of colloidal solution phases are little inferior to them in their ability to introduce order into the chaotic dance of the solutions' molecules. We shall not dwell upon the out-datedness of the genetic information which Shrödinger presented; it is excusable for a physicist to be out of touch with the latest innovations in biological theory. The essence of the matter does not lie here. It appears probable that no physical or physico-chemical phenomena, when taken separately, can provide an adequate explanation for the processes of preserving and decreasing entropy in the organic world (and in cybernetic machines in which it is possible for the amount of information to grow, which is also one expression of increased negative entropy; cyberneticists think that augmenting the amount of information is also possible in natural living bodies).

Not only the properties of the macroscopic solid body, but the combined properties of several aggregate property states – liquid, colloidal, semi-liquid crystals, solid and gaseous – are characteristic of living bodies in this respect. A complex combination of different properties and processes is realized in protoplasm based on its multiphase properties. The division and breakdown into relatively distinct subsystems and components, as we have pointed out, has in general great meaning for the integrity of an organism and for its entire destiny, not only to ensure its dynamic stability but also to ensure adequate rates of adaptive variability and the further self-development of the living system. In addition, the subdivision of the organism is one of the bases for developing a many-sided centralization of functions, this being combined with the preservation and development of its potentials with regard to free changes in its individual components. Helping to increase the heterogeneity of changes and structures, the multiphase nature of living systems is a far more important basis for the rise of negative entropy than the properties of aperiodic crystals.

An ever more important part is played by another expression of ectropism in organic life – the ability of the living body to extract from the environment dispersed and particularly less highly organized materials and energy sources (or to decompose highly organized bodies into their components), so as to transform and concentrate them in themselves, and then to use them as their own organized materials and forces for their active internal and external activity. Shrödinger wrote about the importance of 'nourishing organisms with negative entropy', but this idea, though correct, must be filled out considerably.

Of course, having decisive practical importance is the degree of biological efficacity in the actions of living bodies. This includes the physical and 'technical' efficacy of internal and external actions. To determine merely the simple physical efficacy we must consider, not so much the actions of the individual specimen, as the life of the species in the conditions of the systems of species. But here we will confine ourselves to a preliminary qualitative description of the physical efficacy of the individual animal.

Biophysics has not yet, so far as we know, fulfilled the task of studying one of the most remarkable properties of living bodies which has served as a 'natural premise' for such phenomena in human society as surplus labor, about the significance of which there is no need to speak here. This is the development of the ability to perform a larger total amount of external work than is spent on bearing, developing, and reproducing the organism. With regard to the organism's negative entropy, this is the process of its auto-augmentation. From this point of view life is self-augmenting negative entropy.

The amount of external and internal work which surpasses the biological expenditures on reproducing the organism and species can be called surplus work (in the physical sense of the word). One of the components of this concept is surplus energy. It is clear that these two ideas describe the core of the matter better than the idea of the maximum stream of energy (A. Lotka, after Rashevky), since the biological efficacy need not rise if the increase and intensification of the energy stream is bound up with too great a rise in the expenditures of energy (where there is a low efficiency factor). It is on the phenomenon of surplus external and internal work that is based the possibility of expansive reproduction of populations and species, and in society the possibility of using domesticated animals for work. A basically similar problem has constantly faced the technical sciences, and particularly cybernetics. Certainly no activity performed by domesticated animals and automatic devices can ever be transformed into labor. But they do perform work – external and internal to them – in the physical and more complex technical sense of the word.

Surplus work can be measured. Something which may be one of the most important measuring indicators (parameters in the customary sense of the word) is, we believe, the coefficient of surplus work.

This aggregate indicator contains less complicated factors, such as the physical power and coefficient of useful action. In turn the surplus work coefficient itself becomes one of the subordinate factors in the indicator of over-all biological efficacy and progress (and in application to technical systems, in the indicator of their technical progressiveness and over-all economy).

In the progressive evolution of animals, there is an increase in the amount and complexity of the external obstacles which they must surmount in order to avoid actions that are more remote in space and time, i.e. in order to avoid more remote external dangers.

In this respect the following circumstance is of importance, from the standpoints of both physics and biology. Every individual act of nervous and external activity in the higher animal, and this is in complete accordance with the general laws of physics, expresses the well-known principle of minimal action. But when we analyse fairly large (and usually cyclical) portions of the 'behavior line' (we use this term here simply in the sense of the aggregate of individual behavior acts) we discover something basically different – the behavior line is, for a flourishing population as a whole and for the average cycle, the line of the largest possible total amount of surplus work.

With regard to external bodies, each individual action of an animal is always a movement along the line of least resistance, along the resultant of the external forces present. However, by paying heed only to analogies with the general laws of motion, physics has lost sight of a fact which present-day cybernetics must reckon with from the very start – the fact that every individual act of a living system possesses, in essence, a triple nature.

Most directly it is a movement, a change in the given material system. The specific peculiarities of the particular movement contain the mediated and coded reflection of information about the past history of the species and environment. In addition, the peculiarities of the movement may also reflect, to some degree, its programatic or 'prospective' significance for the future history of the individual, species, and also the environment.

On the basis of biological rather than physical laws, there has developed a property which is paradoxical from the standpoint of physics – movement along the line of the greatest total amount of surplus labor. But we think it possible to show that this phenomenon includes a higher development of that general physical principle of movement 'along the line of least resistance' – movement along the line of the smallest total sum of external resistances.

The ability to surmount ever greater obstacles, to defy unfavorable external circumstances in an ever larger arena of environmental conditions, expending at first more energy on this than would be required for passive defense from harmful influences – with the goal of curtailing the total amount of energy expenditures and improving the chances for a victorious outcome in the struggle for life – this ability takes on greater and greater importance as the progressive evolution of animals proceeds.

There seems no doubt that this important feature of living systems can be studied quantitatively and quite objectively in mutual connexion with the 'calculation of progress'.

8 G. Sommerhoff

The Abstract Characteristics of Living Systems

This paper is published for the first time, in this volume.

Preface

If the concept of organization is of such importance as it appears to be, it is something of a scandal that biologists have not yet begun to take it seriously but should have to confess that we have no adequate conception of it (J. H. Woodger).

The physico-chemical picture of the living organism is only half the truth. The missing half concerns the nature of the organizational relationships that make the behaviour of obviously living systems uniquely different from that of obviously non-living systems and give it a teleological quality not found elsewhere.

In many ways this is the more important half. For here lie the differences between life and death, and between higher and lower forms of life as they affect us most. Here too lies the key to the true function of the living brain, for this function is nothing if not an organizing and teleological one.

The distinctive organization of living systems shows itself in the goal-directedness and directiveness of their activities. Aristotle's 'absence of haphazard and conduciveness of everything to an end' are not two separate criteria of life. We infer the former from the latter. The lesson to be learnt from cybernetics is that we are dealing here with objective system properties which, in principle at least, are susceptible to mathematical analysis.

But they are very abstract properties, and to get a clear conception of the distinctive nature of vital organization in general and the distinctive functions of the brain in particular we must above all acquire the ability to think of living systems in sufficiently abstract terms. We also require the *purest form of behaviourism.* That is to say, we must start by thinking of the living organism as no more than a 'black box' which reacts to its environment in certain observable ways, and then analyse mathematically *what* it does, without preconceived ideas of *how* it does it.

Biologists, psychologists and neurologists have been all too slow to realize the importance of abstract and accurate analysis along these lines. They are still content to leave it to the cyberneticists. But cybernetics has its origin in control engineering, and so it is not surprising that the biological sciences have to pay for their omissions by an undue servitude to engineering concepts. That is why the writer continues to press for independent pure research in this field – for what he has called *analytical biology.*

1. Introduction

1.1. Even if we knew down to the last molecular detail what goes on inside a living organism, we should still be up against the fact that a living system is an organized whole which by virtue of the distinctive nature of its organization shows unique forms of behaviour which must be studied and understood at their own level, for the significance of all living things depends on this.

The most distinctive character of the behaviour of higher organisms is its goal-directedness, its apparent purposiveness. In fact, it is largely through this apparently teleogical nature of their activities that living organisms betray their exceptional organization. And their position on the 'scale of life' is largely determined by the degree to which they possess those characteristics.

The task of analytical biology is to concern itself with the systematic study of these abstract characteristics of life, and with the precise analysis of the ordering relations in space and time on which they depend. To understand this particular aspect of life is important not only for the biologist: it profoundly concerns also the psychologist, philosopher, and automation engineer.

The first requirement of any systematic study of these abstract characteristics of living systems is to find a set of exact concepts which have the power to describe and analyse them with mathematical precision, and at the same time stand in a definite relation to the concepts used in the physicochemical description of life. The intrinsic unity of life must not be lost.

The object of this reading is to outline such a basic set of concepts. It is not concerned with the formulation of new hypotheses or the construction of theories – only with the necessary groundwork, with the conceptual tools. It is concerned not with our powers to predict but with our powers to define and define accurately.

1.2. Many animal activities are patently goal-directed in a way which makes it clear that we are dealing here with an objective system-property – a property, moreover, which we can assume to be compatible with the basic laws of physics and chemistry. We can assume this because we know of servo-mechanisms and automata that are capable of essentially similar types of behaviour.

The movements of a chick pecking at a grain, a rabbit digging its burrow, a pike chasing its prey, a bee homing on its hive, are the sort of examples one could quote of activities that can be paralleled to a significant extent by known forms of automatic control as used in industry, aircraft, guided missiles, etc. At a different level one can draw examples from the various regulating mechanisms that maintain the normal state of the organism, for example the mechanisms that control our body temperature, or those that regulate the carbon dioxide or glucose concentration in the blood.

Many other vital activities or processes are goal-directed in the same objective sense, although we have no obvious mechanical analogues for them: embryonic growth, maturation, insect metamorphosis, regeneration, and others.

Beyond this whole field of activities which are goal-directed in the full sense of the word, and where one can name a specific goal towards which the activity is directed, there is an even wider field of organic activities which are directive in, perhaps, a weaker sense. Here the 'goal' is not so much the occurrence of a specific event, situation, or configuration – or their restoration – but rather the attainment of some general improvement in the organism's capabilities or performance. This may be an improvement in some anatomical structure; or it may be the better adaption of a behaviour pattern; or perhaps the skill with which an animal can perform a certain task. Evolution is directive in this sense. Through natural selection it leads to modifications of the organism or its behaviour patterns which favour the survival or reproduction of the individual. Some forms of learning come into the same category. Here again we are dealing with objective system properties which we can assume to be compatible with the laws of physics and chemistry. We know the general 'mechanism' of natural selection and we can build automatic machines which 'learn' in the sense that, evaluating the results of previous activities, they can improve their performance as they go along.

1.3. Frequently, indeed almost invariably, various different forms of goal-directedness or directiveness are all present and intimately inter-related in the same biological situation. This is a common source of confusion. In particular, one of the most distinctive characteristics of the living organism is the hierarchical manner in which the goals of the various part-activities (or activities of its parts) are inter-related and integrated. In a typical case each part-activity has a proximate and transient goal, which is itself subservient to the less transient and less proximate goal of the action as a whole, which in turn is subservient to the goal of the behaviour patterns as such, and so on. Or, it may be the goal of one activity to establish or maintain the conditions under which another goal-directed activity can properly take its course; while another activity in turn establishes or maintains the prerequisites for the first, etc., all of this involving structures and mechanisms which are themselves the product of directive developments. There is nothing to parallel this co-ordination and integration in the world of machines and automata.

These are organizational relationships, and this type of coordination and integration is of the essence of vital organization. We are dealing here with a higher, teleological, type of order that is quite distinct, for instance, from the mere geometrical order that distinguishes the molecular structure of solids and forms the subject of classical entropy studies. The ordering relations here are in time as well as in space.

In greater or lesser degree this teleological order is a primary characteristic of all obviously living systems. More than any other property it is the one that sets them apart among the objects of our environment and determines our reactions to them. It is of the essence of the connotation of life.

One of the major confusions of our time and one that distorts all our thinking about the phenomenon of life as such is the confusion between the connotation and denotation of the concept of life. I may illuminate this with a different example. 'Featherless biped' covers accurately the denotation of the concept of man. That is to say the class of known humans is co-extensive with the class of known featherless bipeds. But the attribute of being a featherless biped bears no relation to the connotation of the concept of man. If a sudden genetic mutation were to result in some of us producing feathered (but in other respects normal) offspring we would still class the offspring as humans in the accepted sense of the word.

Similarly, the denotation of the concept of life may be adequately covered by the conjunction of a number of empirical attributes such as self-reproduction, growth, metabolism, etc. But again these attributes bear no relation to the connotation of life. And this is immensely important for two reasons. In the first place all our human reactions to the phenomenon of life are in response to the attributes that the concept connotes and not to the attributes that make up the denotation listed above. When our pet dog dies we do not mourn the irreversible loss of its power of self-reproduction, growth, metabolism, respiration, and the like – we mourn the irreversible collapse of the organized and goal-minded activities that made the dog the companion he was. Secondly, just as we must expect the science of man to be more than the science of the property of being featherless and a biped, we must expect a science of life to be more than a science of the attributes that happen to cover the denotation of the concept. And yet that is all modern biological theory amounts to. There exists as yet no pure theoretical biology.

But the characteristic teleological order of living systems is important in other respects, too. It directly concerns the theoretical problems that face the ethologist, psychologist, and neurophysiologist and workers in other biological sciences. It also concerns the engineer who is interested in brain-like mechanisms. The central nervous system more than any other part of higher organisms is responsible for introducing teleological order into the activities of the system. And just as much as the biologist needs exact concepts and an exact analysis to be able to deal with these properties in whatever form he meets them, the engineer needs such concepts for his design studies on brain-like mechanisms or for simulating on a computer the most typical characteristics of brain-like behaviour – either again as design studies or for testing hypotheses about the possible mechanisms that may be operative in the living brain. Much analytical work remains to be done before the most truly significant characteristics of the brain's activity can be grasped in physical or engineering terms. Until then the engineer remains limited to picking the bits of function that he can grasp and copying these on a computer or hardware. But the properties of the whole are apt to be lost in this process of fragmentation.

1.4. It is only in recent years that these characteristic properties of living systems have come to be looked upon as objective system-properties that require an independent abstract analysis within the framework of a general and exact systems research. They are elusive properties which, partly because of the vague and ambiguous terms in which they were discussed, have been the subject of much inconclusive speculation in the history of biology. But the elusive must be grasped and bounds must be set to the vague. Many of the relations involved are too intricate to yield to a mere verbal analysis. Here only a formal analysis can help. From what has been said above it is evident that a formal analysis of goal-directedness and directiveness in organic processes, and of the hierarchical relations that we always find here, has an essential part to play in this respect.

The analysis must be general and exact. In the present context the former means that we shall primarily be concerned with the general aspect of *what* a system does and not *how* it does it. The details of the mechanism by which this type or that type of goal-directed behaviour is accomplished, are not our primary concern. And to say that the analysis must be exact means that it must be driven to the point at which it can be expressed in terms of mathematical relations between physical variables. Some systems show goal-directed behaviour, others do not. What functional relations (in the mathematical sense) exist between the physical variables of the former that are absent in the latter? That really is the basic question to start with.

1.5. The working biologist may often refer explicitly to the goal-directedness or directiveness of biological activities. But many of his key concepts imply it; concepts like adaptation, coordination, control, regulation, learning, maturation, instinct, drive, etc. At present most of these concepts suffer from a degree of ambiguity and uncertain meanings which hamper the progress of theoretical biology. This is also true of psychology and ethology, as the history of behaviour theory well shows.

In a very limited sense this impasse has been overcome by the growth of servo-engineering. Although the servo-engineer deals with quite different structures, a number of his general concepts can profitably be applied to the behaviour of the living organism. And since many servo-mechanisms produce some type of goal-directed behaviour it is not surprising that some of the general concepts of servo theory can illuminate parallel forms of goal-directed behaviour in living systems – concepts like error control and feedback loops, for instance. But there are many limits to this. The goal of the activity of a servo-mechanism is set by an external command. This command provides the reference signal from which the error signals are derived to which the feedback mechanism responds with the necessary corrections. These concepts fail us when, as in so many biological cases, the commands are merely transient ones which themselves form part of a goal-directed activity of a higher order. In other words, they fail us when we face typically integrated forms of biological order. They also fail us in other respects – in many biological activities there is nothing that could be identified as a reference signal or error signal in the technical sense without unduly stretching these concepts, or without introducing fictitious quantities which are apt to confuse rather than simplify the matter. It is difficult to see, for instance, in what exact sense one can think of reference signals in such goal-directed processes as embryonic growth, maturation, learning, evolution, etc. – or even in such simple cases as a rat taking the correct turning at a choice point of a maze. And, as we shall see, the concept of error signal is also beset with difficulties.

Without denying the value of some engineering concepts in certain cases of regulative activities, their application does not diminish the importance of developing concepts which with the same accuracy manage to capture the more truly distinctive characteristics of the overt behaviour of living systems, particularly at the upper end of the phylogenetic scale.

1.6. The psychologist is preoccupied with the mind rather than the brain. That means that he is interested in the subjective rather than the objective side of goal-directed behaviour. We know from introspection that in some cases goal-directed behaviour (in the objective sense) is achieved through the instrumentality of conscious thought processes – through the conscious striving after some future goal which is retained as some kind of idea or image. But since we also know that an action does not necessarily have to be consciously purposed in order to be goal-directed in the objective sense of the term, it is advisable to maintain a strict distinction between 'purposive' in the subjective sense and 'goal-directed' in the objective sense. The present paper deals only with the latter. This does not preclude the possibility of making a transition to the subjective aspects of behaviour at a later stage. But it could not be the other way round.

Any exact science must start with what is publicly observable. The essence of scientific concepts is that they are definable in terms of public operations and observations. That is the very secret of their power. The exact sciences, therefore, cannot proceed from things of which we only have private awareness. They must start with objective and public events. But there is no reason why it should not be possible at a later stage to interpret subjective events in terms of such objective events.

2. Some Fundamental Physical Concepts

2.1. Throughout our analysis, the organism and its environment will be treated as a dynamical physical system. In the context of the discussion this system is a macroscopic one and the analysis can therefore be based on the conceptual framework of classical physics. The following is a brief outline of some of the concepts that will be used later on.

An isolated physical system in the classical sense is *state-determined.* This means, *inter alia*, that the state of the system at any instant can be represented as a single-valued function of the *initial state* and the *time* co-ordinate. For instance, the velocity and displacement of a simple pendulum at any instant can be expressed as a single-valued function of the velocity and displacement with which the pendulum was released and the time that has elapsed since.

Whenever possible this determinateness is assumed by the working biologist, as it forms the only basis for single-valued predictions. It is also implied in the concept of 'mechanism' and must therefore be assumed whenever we look for mechanisms of one kind or another to explain some form of animal behaviour.

In the context of the present investigations the organism itself cannot be treated as an isolated physical system, because it interacts with its environment, and it is precisely with these interactions that we shall be concerned. But the conjoint system of organism plus environment may be treated as isolated and, therefore, state-determined in the above sense. By 'environment' we shall understand those surrounding conditions which affect the organism or the results of its actions.

2.2. The physicist deals with the complexity of concrete objects and events by inventing abstract models or representations in which certain factors are singled out and the others discarded as irrelevant. The real pendulum, for instance, may be represented by the familiar plane-mathematical pendulum consisting of a mass-point at the end of a weightless thread. In this manner abstraction is made from irrelevant factors such as the colour of the pendulum or the shape of its bob. In such theoretical representations the state of the system at any given instant is defined by specifying the values for that instant of a selected set of variables. These may be called the *state variables*. Their number depends, of course, on the degree of abstraction and simplification that are made in the model and on the *conditions* that are assumed to be constant for the purpose of the investigation in hand. In many cases, for instance, the length of the pendulum would be treated as a constant parameter, and would not therefore be one of the state variables. The choice of the theoretical model must invariably depend on the context in which it is used.

The variables of a physical system have various kinds of mutual interdependence and independence. One type of independence is particularly important in the treatment of our subject and must be singled out:

Definition: Two variables are *orthogonal* if the value of the one at any instant of time does not determine the value of the other *for the same instant*.

Example 1. The velocity and the displacement of a pendulum are orthogonal variables, because the value of the one does not determine the value of the other for the same instant. But the displacement and the acceleration are not epistemically independent because the value of the displacement determines the value of the acceleration for the same instant.

Example 2. In a servo-mechanism the input (command signal) and the output are orthogonal, because owing to the inevitable exponential lag (which may be supplemented by a transport lag as well), the value of the one does not determine the value of the other for the same instant. On the other hand, in the absence of a transport lag, the input and the rate of change of the output may not be epistemically independent.

If two variables are orthogonal, this does not of course preclude that the value of the one at any instant may influence *subsequent* values of the other.

One sense in which the concept of orthogonality is important concerns the assumptions that can be made about the possible initial stages of a system. In a deterministic system the succession of states through which the system passes is determined by the initial state. We generally assume that we can postulate as an initial state of the system any random combination of the system's state variables. But obviously this can only be true to the extent that the selected set of state variables are mutually orthogonal in the above sense.

Unless explicit mention is made to the contrary, we shall always assume that the state variables of the organism-plus-environment system are orthogonal in this sense.

2.3. We shall assume throughout that the organism-plus-environment system is state-determined, so that the current state of the system can be expressed in terms of single-valued functions of the initial state and the time. To illustrate a simple system of this kind, consider an object moving along a straight line with a uniform acceleration a. The position of the object at a time t is then given by

$$x_t = \frac{at^2}{2} + v_0 t + x_0$$

where v_0 and x_0 are the initial velocity and position.

With v_0 and x_0 given, this may be read as giving x as a function of the time t:

$$x = f(t) \qquad (1)$$

Or, with the time given, say $t = t_1$, it may be taken to express x at t_1 as a function of the initial velocity and position:

$$x_1 = F(x_0, v_0). \tag{2}$$

In the sequel, functions of the last type will prove to play a large part in our analysis of goal-directed behaviour.

Owing to the complexity of the living organism it will not generally be possible to give explicit formulae for any of the functions concerned. Some functions may be discontinuous functions or step functions. Other functions may have an argument with a very restricted and finite domain. For instance, some variables encountered in the study of animal behaviour are not intrinsically quantitative. An example would be a rat turning either right or left at a choice point in a maze. A variable of this kind may be given a numerical expression by assigning (say) the number 1 to the case when the rat turns right and the number 2 to the case when it turns left. This is known as assigning a *characteristic function* to the variable. The result is a binary variable.

None of this need affect the rigour of the analysis. The majority of statements in mathematics, after all, are about functions for which no formula is given.

The functional symbols (f and F above) may always be interpreted either as an abbreviation for a known formula (as above) *or* as a symbol for a *correspondence* or *mapping* between two sets of objects. And the concept of function in this sense is so general that the objects may be objects of any kind. For instance, each object of the set may itself be a sequence of several numbers. In this sense a function of several independent variables u, v, w, may always be treated as a function of a single variable $\mathbf{z}$. $\mathbf{z}$ then represents the sequence (u, v, w) and is called a *vector:* u, v, w, are its *components*. A function linking two vectors is equivalent to a *set* of functions linking their components.

2.4. When the state of a physical system is described in terms of N epistemically independent state variables, it may be represented as a point in an N-dimensional space. The N numbers defining the state of the system are then the N components of the position vector of the *representative point*. And the behaviour of the system in time is reflected in the movement of the representative point in the *phase space*. As it moves it traces out the *line of behaviour* of the system. In a state-determined system every possible initial state determines a single line of behaviour only. And although different lines of behaviour may at one point or another fuse, they can never bifurcate.

For any point of time t, the representative vector $\mathbf{p}_t$ is a single-valued function of t and of the initial vector $\mathbf{p}_0$. That is to say, the value of each component of $\mathbf{p}_t$ is a single-valued function of t and of one or more components of $\mathbf{p}_0$.

2.5. There is a subtle difference between the case when a state-variable is presented as a function of the time t (2.3, equation **1**) and when it is presented as a function of the initial state of the system (2.3, equation **2**).

In the first case the function describes the *behaviour* of the system in time. But in the second case the function describes how the state of the system *would* have differed from the actual one, *if* the system had started from a different initial position. Now the function describes a *property* of the system. And although the description is in terms of *hypothetical* variations of the actual situation, it represents none the less an *objective* property of the *actual* system.

This is a common case in physics. For instance, an object is said to be in a state of stable equilibrium if a hypothetical (virtual) displacement entails an increase in the potential energy. Whether this displacement can be carried out in practice is immaterial. It must merely be theoretically possible, i.e. compatible with the degrees of freedom of the system. Yet, despite the purely hypothetical character of the displacement, stability is regarded as a real and objective system property. Indeed, it can be said that in the last analysis all classical physical properties are of this kind. The temperature of a water-bath is defined in terms of the reading a thermometer *would* give *if* it were inserted into the bath. The water does not cease being hot when the thermometer is removed.

We shall find that the system properties studied in these papers are similar, in that they can be defined only in terms of hypothetical variations of a given situation.

3. Preliminary Analysis of Goal-Directed Behaviour

3.1. To discover what the exact relationships are that distinguish a system showing goal-directed behaviour from one that does not – and so to give a precise meaning to this concept in the language of mathematics and physics – is not a simple task of translating one language into another. For one thing, we are dealing here with a term which as yet has no uniform usage. The first step therefore must be to clarify further the sense in which the concept is to be understood here.

It has already been said that it must be understood in an objective sense in which it does not presuppose the existence of 'purposes' in the subjective sense, and in which it can be applied to machines as well as to living systems.

On the other hand, goal-seeking is not to be understood in such a generalized and trivial sense that, for instance, a falling stone would become an instance of goal-directed behaviour. In such a trivial interpretation our concepts would fail to capture those very distinctive characteristics of organic activities that we are interested in. We must be guided in this respect by the ultimate purpose of the investigation in hand.

Examples of goal-directed behaviour in our sense, therefore, would be a chick pecking at a grain – but not the blind responses of a clockwork imitation; an aeroplane guided by an autopilot – but not the unavoidable course taken by a truck guided by rails; a bird building a nest – but not the unavoidable effect of water hollowing out a cavern; men digging away a mountain – but not nature throwing up one. Directive, in our sense, would be the evolutionary adaptation of a plant or animal species to its environment – but not the blind evolution of the solar system. Excluded, too, would be all instances of animal activity in which the biological significant result is merely an accidental or fortuitous one.

This contrast of the results of goal-directed activities with what is unavoidable on the one hand, and accidental on the other, is significant and illuminating. Although we shall not be in a position to clarify the exact relationships between these concepts until a much later stage, this juxtaposition provides a convenient starting point.

3.2. To follow this up, let us take a concrete example and examine what operational procedures we would adopt to verify the existence of goal-directed behaviour. As we want to concentrate on *what* the thing does without being distracted by *how* it comes to do it, we may borrow for this purpose the familiar engineering concept of a 'black box'. That is to say, an entity about whose structure nothing is assumed to be known.

Suppose then that on a table I find a mysterious black box fitted with a switch. And a couple of feet away there is a red ball. I press the switch. After a short buzz a black ball is suddenly shot out of the box and hits the red one. I am now faced with the following possibilities:

i. That the hit was accidental.
ii. That the switch merely released the ball and that the hit was unavoidable owing to, say, a guiding groove in the table that I had not noticed.
iii. That the hit was the result of some goal-directed activity either (a) on the part of some servo-mechanism in the box that scanned the table and then aimed the black ball at the red one, or (b) on the part of the person who placed the objects on the table and may have oriented the box.

I can eliminate ii by a physical inspection of the table and the surroundings of the box or ball. To eliminate i and iiib, and so to establish iiia, I would obviously proceed by placing the red ball in a new position, *out of the original line of fire*, and then operate the switch again. If again a black ball shot out and hit the red one I would begin to suspect that these hits were not accidental; and if repetition of the procedure produced the same results I would conclude that the box contains a servo-mechanism producing some sort of goal-directed action. In other words, *the test procedure is to vary the circumstances in a manner that calls for a modification of the action and then to test whether successive repetitions of the action show the required modifications.*

The example shows that the concept of goal-directedness does not merely relate to a single concrete action, but *in addition* to either of two things:

i. how the box acts *when* the circumstances are varied in a certain way and the action *is* repeated, *or*

ii. how the box *would have* acted *if* the circumstances had differed in certain ways.

We must now decide between i and ii. That is, we must decide whether the concept relates i to a *sequence of actual occurrences* or ii to a *single* actual occurrence and a *set of hypothetical variants* (the actual repetition being merely necessary as a test procedure).

To decide this, imagine an exactly similar box except that this time the box contains only a single ball, in an inaccessible magazine. The box fires and hits. I would then not be able to verify whether the hit was accidental or the result of goal-directed activity. But obviously I could nevertheless entertain the second possibility as a meaningful hypothesis. And if there were a number of such boxes I could establish the likely truth of the hypothesis statistically by firing them all in turn. This clearly establishes ii as the correct interpretation. In other words, a goal-directed action is one which not only leads to the 'goal event' in the actual circumstances, but in addition is determined in such a way that *if* the circumstances *had* differed in a manner calling for a modification of the action, the action *would* have shown the required modification. All within obvious limits, of course.

It is seen from this that, similar to the physical properties mentioned in para. 2.5, goal-directedness is a property which in the last analysis can only be defined in terms of hypothetical variations of the actual situation. It is none the less a real property of the system. But it is a property of the mode of operation of the mechanism rather than of its output. What it really says is that the mechanism *is* such that not only *did* it act appropriately in the given circumstances, but also *would* have acted appropriately *if* the circumstances had differed from the actual ones in a way that would have called for a modification of the action.

This is easily seen to be true for all cases of objectively goal-directed behaviour. When a bird pecks at a grain we know that at the time it is so conditioned that if the grain had been in a different position, the bird would have modified its movements accordingly. A player stops a ball with his foot. Was this a 'fluke'? No, not if we have reason to believe that if the ball had taken a somewhat different course (again within obvious limits) the player, too, would have placed his foot in a different – and once again appropriate – position.

3.3. Provisionally we may sum up this point as follows:

If in an environment *E* an action *A* is directed towards a goal *G*, this implies:

i. That *there exists a set V of hypothetical variations of the environment such that each member of V requires a specific modification of the action A if the goal-event G is to result*, and
ii. The organisms or machine at the time is so conditioned that *if any of these variants had in fact been the case, the action A would have shown the required modification.*

'Specific modification' here means that no two members of the set V require the same modification. In the example of our 'black box' moving the target ball to positions *p* or *q* is a set of variations requiring a specific modification of the action, but moving it to *s* or *t* is not.

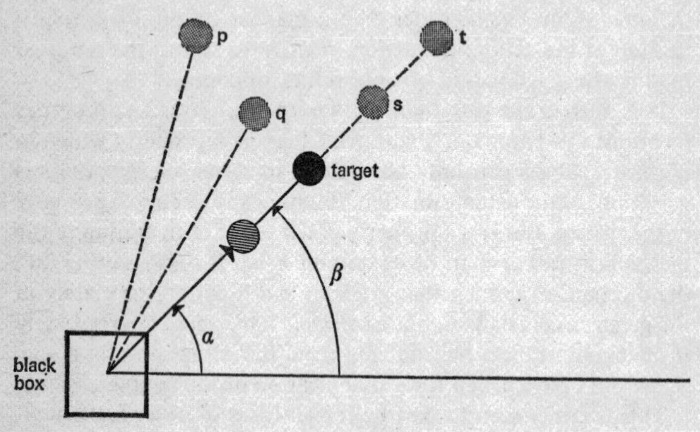

Figure 1

Goal-directedness is a matter of *degree*. And in a sense it seems a reasonable interpretation to say that this degree depends on the magnitude of the set V for which the conditions specified under i and ii above hold good.

But the analysis must be taken further. We still lack the precision we require. For instance, there is an apparent contradiction between statements i and ii. In i *E* and *A* are thought of as *independent* variables and *G* as a function of these. What it says in effect is that certain specific combinations of *A* and *E* result in *G* and other combinations do not. But in ii *A* and *E* are thought of as *dependent* variables, the action *A* being thought of as a function of the environment *E*.

This apparent contradiction is due to the fact that i and ii refer to different 'time-slices' of the environment. Taken together both statements relate to the behaviour of the system over a certain stretch of time – beginning with the stimuli that release the action and determine the form it takes, and terminating with the occurrence of *G*. Now the environmental variations referred to in i relate to the 'time-slice' of the environment that decides the result of the action, i.e. a time-slice that *coincides* with the action. Whereas those referred to in ii relate to the time-slice of the environment to which the animal or machine reacts, i.e. to a time-slice that *precedes* the action (or the relevant parts of the action). And, to complicate matters, these two time-slices may overlap.

A more detailed treatment is therefore required in which all the variables involved are set out as functions of the time variable. The situation must be analysed in terms of the transition of the system from momentary state to momentary state. There is no other way of sorting out the warp and weft of dependencies and independencies that are tacitly implied in the concept of goal-directed behaviour.

4. Final Analysis of Goal-Directed Behaviour

4.1. The analysis of the last section has brought out one essential ingredient of the concept of goal-directedness. This was formulated in i and ii on p. 162. We must now complete the picture and at the same time translate our findings into a language which makes it quite clear what they mean in terms of a physical description of the system in which its variables are regarded as functions of the time variable and the initial state (see section 2).

In all, the concept of goal-directedness implies that the relations between the action, the environment, and the goal-event satisfy all of *three criteria*. These will be discussed in turn below.

Criterion I

In goal-directed behaviour the action itself is not a sufficient condition for the occurrence of the goal-event. In all cases the occurrence of the goal-event is conditional on the action having a specific relation to the environment. More specifically, the situation is always one in which it is a necessary condition for the occurrence of the goal-event that one or more action variables stand in a particular relation to one or more environmental variables.

So if an action is directed towards a goal G this implies that there exists at least one point of time t_k during the action and at least one variable $\mathbf{a}$ associated with the action, and at least one variable $\mathbf{e}$ associated with the environment, such that at t_k it is a necessary condition for the occurrence of the goal-event G that $\mathbf{a}_k$ has specific relation to $\mathbf{e}_k$, i.e. that it satisfies some particular function

$$F(\mathbf{a}_k, \mathbf{e}_k) = 0 \qquad \textbf{(1)}$$

whatever that function may be. (The term 'variable' in this context is, of course, to be understood in the generalized sense in which, for instance, it may denote a set of variables or a vector variable.)

Thus in the example of the black box (Figure 1) it is a necessary condition for a hit on the target that at the time of firing (t_k) the angle α should equal the angle β. So $F(\mathbf{a}_k, \mathbf{e}_k) = 0$ here takes the form $\alpha_k - \beta_k = 0$.

In the example of the bird pecking at a grain two cases must be distinguished. The action may be of a stereotyped nature so that once released it takes a fixed course. Or the pecking action may be continuously modifiable and error controlled as in a servomechanism with feedback. But our argument is not affected by this difference.

In either case at *some* stage, *some* action variable must stand in a particular relation to the position of the grain if the action is to be successful. In fact, the very concept of error control implies that the movement stands in a particular relation to the error, and the error, of course, is a function of both the position of the beak and the position of the grain. A necessary condition for success of the action is, for example, that at some stage at any rate during the movement the rate-of-change vector has the opposite direction as the error vector. And there are other necessary conditions, but the precise form the respective functions take depend on the constants and restraints of the system and in what degree of detail the system is specified.

Criterion II

Secondly, the concept of goal-directedness implies that the variables **a** and **e** in the function *F*, introduced above, are mutually orthogonal (p. 164). For the implication always is that $F(\mathbf{a}_k, \mathbf{e}_k) = 0$ is a condition that must be *brought about* by the mechanism involved. It is not implied by the axioms of the system. Indeed, random combinations of **a** and **e** are conceivable initial states of the system.

The point has already emerged at various stages of the discussion. This is, for instance, what distinguishes the case of the truck guided along rails (which is not goal-directed behaviour in our sense, see p. 159) from the aeroplane guided to its destination by an autopilot (which is).

It is also particularly obvious in the example of the black box, (Figure 1). The servo-mechanism in the box adjusts α to β. This very notion implies that $\alpha_k \neq \beta_k$ is a possible initial state of the system. If this were not the case, no servo-mechanism would be required.

Similarly, in the example of the bird pecking at a grain, states in which the bird pecks at something else or merely executes random movements are conceivable states of the system.

That in a servo-mechanism, random combinations of input and output are conceivable, and hence possible initial states of the system, was already discussed in section 2.

Criterion III

Finally, the third implication of goal-directedness is the one that was discussed at length in section 3. The concept implies that the animal or machine produces the appropriate action (and satisfied $F(\mathbf{a}_k, \mathbf{e}_k) = 0$ not only under the actual environmental conditions, but also that it would have produced an appropriately modified action under a variety of alternative circumstances each requiring a specific modification of the action.

In saying this we are making a statement about the behaviour of the system during a certain slab of time. This time-slab comprises the incoming signals or stimuli, the 'computing stage', the action and, finally, the consequences of the action. We may call the starting point of this time-slab the 'initial time' t_0, and the state of the system at t_0 the 'initial state'. The point of time t_0 may be chosen arbitrarily provided the resulting time-slab comprises all the above elements.

Now, if physical determinacy is assumed the succession of states through which the system passes is determined by the initial state (section 2). If in the present context, therefore, we are talking about the behaviour of the system under alternative environmental circumstances, we are saying something about how it would have behaved if it had started from alternative initial states. And what, in fact, we are saying is that not only does the action satisfy $F(\mathbf{a}_k, \mathbf{e}_k) = 0$ for the actual value of $\mathbf{e}_k$, but in addition:

i. That there exists a set of alternative initial states each of which would have led to a different value of $\mathbf{e}_k$, and would have required a different value of $\mathbf{a}_k$ if $F(\mathbf{a}_k, \mathbf{e}_k) = 0$ is to be satisfied and G to occur.
ii. The organism or machine at the time is so conditioned that in each of these cases it would in fact have produced the appropriate value of $\mathbf{a}_k$, and so would have satisfied the above equation.

All this obviously implies that this set of alternative initial states must differ in a variable of which both $\mathbf{e}_k$ and $\mathbf{a}_k$ are functions. Let this be an environmental variable $\mathbf{u}$ and let $\mathbf{u}_0$ denote the value $\mathbf{u}$ at t_0. Formally, therefore:

$$\mathbf{e}_k = E_k(\mathbf{u}_0)$$
$$\mathbf{a}_k = A_k(\mathbf{u}_0).$$

($\mathbf{a}_k$ comes to be a function of $\mathbf{u}_0$, of course, by virtue of the way the organism or machine reacts to the environment.)

And the statements i and ii further imply that there exists a set S_0 of values of $\mathbf{u}_0$ such that:

(a) S_0 has at least two members; and

(b) no two members of S_0 result in the same $\mathbf{e}_k$ or the same $\mathbf{a}_k$, i.e. the functions E_k and A_k are bi-unique for all $\mathbf{u}_0$ in S_0; and

(c) the action is shaped in such a way that for all values of

$\mathbf{u}_0$ in S_0, $\mathbf{a}_k$ satisfies $F(\mathbf{a}_k, \mathbf{e}_k) = 0$; that is to say $A_k(\mathbf{u}_0)$ satisfies $F[A_k(\mathbf{u}_0), E_k(\mathbf{u})] = 0$.

In the example of the black box, t_k is the time of firing. $F(\mathbf{a}_k, \mathbf{e}_k) = 0$ takes the simple form $\alpha_k - \beta_k = 0$; as the target is stationary $\mathbf{e}_k = F(\mathbf{u}_0)$ takes the simple form $\beta_k = \beta_0$; and the requirement $F(A_k(\mathbf{u}_0), E_k(\mathbf{u}_0)) = 0$ gives $A_k(\beta_0) - \beta_0 = 0$, and so yields that to make it a goal-directed action the servo-mechanism in the box must establish $\alpha_k = A_k(\beta_0) = \beta_0$ as, of course, we know it must.

In all this analysis the position of the observer (in time) is irrelevant. All the above statements refer to the objective properties of the system during a given slab of time, beginning at t_0. This slab of time may belong to the past, to the present or, it may be imagined, to the future. The only case that might conceivably give rise to confusion is when an observer says of the action of an animal or machine that it is goal-directed, and says so at a time when the action is not yet completed. So part of the relevant time-slab belongs to his past and part to the future. But the *meaning* of the statement is not thereby affected – no more than would be the meaning of the statement 'that robin is building a nest' if it is made before the bird has finished the job.

Summary of analysis

4.2. This completes the main analysis of the fundamental ingredients of the concept of goal-directed behaviour. They may be brought together in the following summary. *If an action is directed towards a goal G, then the following relations exist in respect of an arbitrarily chosen initial point of time t_0, provided this precedes the action by a sufficient interval:*

i. There exist at least one point of time t_k during the action, two variables $\mathbf{e}$ and $\mathbf{a}$ defined on the environment and the action respectively, and a function F, such that

$$F(\mathbf{a}_k, \mathbf{e}_k) = 0 \qquad \mathbf{1}$$

is a necessary condition for the subsequent occurrence of G.

ii. $\mathbf{a}_k$ and $\mathbf{e}_k$ are mutually orthogonal, i.e. $F(\mathbf{a}_k, \mathbf{e}_k) = 0$ is possible.

iii. The action is such a function of the environment that there exist at least one variable $\mathbf{u}_0$ defined on the initial state of the system, and a set S_0 of values $\mathbf{u}_0$ (having at least two members), such that if

$$\mathbf{e}_k = E_k(\mathbf{u}_0) \qquad 2$$

and

$$\mathbf{a}_k = A_k(\mathbf{u}_0) \qquad 3$$

then for all $\mathbf{u}_0$ in S_0, A_k $(\mathbf{u}_0)$ and $E_k(\mathbf{u}_0)$ are bi-unique (one–one) and $A_k(\mathbf{u}_0)$ satisfies

$$F[A(\mathbf{u}_0), E_k(\mathbf{u}_0)] = 0. \qquad 4$$

By 'system' in this context is understood the organism-plus-environment system and, again in the present context, **u** is an environmental variable.

These are the characteristic relations we have been looking for: the functional relations that distinguish a system showing objective goal-directed behaviour from one that does not.

The above formulation is very general. All variables may be generalized variables and all functions may be either continuous or discontinuous.

A special and important case arises if all functions are differentiable and $\mathbf{u}_0$ is a continuous variable within S_0. For in that case equation 4 may be reduced to

$$\frac{\partial F}{\partial A}\frac{\partial A}{\partial \mathbf{u}_0} + \frac{\partial F}{\partial E}\frac{\partial E}{\partial \mathbf{u}_0} = 0 \qquad 5$$

but not identically, for all $\mathbf{u}_0$ in S_0.

This equation is illuminating because it shows particularly clearly that we are dealing here with genuine *physical* system-properties.

It is also illuminating in that it gives the clearest formal expression to the essence of goal-directed activity: viz. that the occurrence of the goal-event G is *invariant* in respect of certain initial state variables ($\mathbf{u}_0$) *despite* the fact that G depends on action factors and environment factors that are *not* invariant in respect of $\mathbf{u}_0$. The invariance of G being due to the fact that the transitional effects of changes in $\mathbf{u}_0$ mutually compensate, so to speak.

Degrees of goal-directedness

4.3. Can a precise meaning be given to the idea of different degrees of goal-directedness? The following illustrates one way in which this may be done.

It is reasonable to assume that the idea in a sense relates to the variety of environmental conditions to which the organism or machine responds with an appropriate reaction. That is to say, it refers to the largest set S_0 for which the relations given in section 4.2 hold good. We may also assume that it implies a comparison of S_0 with the set of all possible variations of $\mathbf{u}_0$. But we must also require that in a case of an action being purely accidentally successful, the degree of goal-directedness is shown to be zero.

Let $\bar{S}_0$ be the largest set S_0 satisfying the relations given in section 4.2; let V_0 be defined as $\bar{S}_0$ minus the actual value of $\mathbf{u}_0$; and let U_0 be defined as the set of all possible values of $\mathbf{u}_0$. Finally, assume that a mathematical measure can be assigned to these sets. The degree of goal-directedness may then be defined by the ratio $q(G; \mathbf{a}_k, \mathbf{e}_k; \mathbf{u}_0) = MV_0/MU_0$, where MV_0 and MU_0 are the measures of V_0 and U_0 respectively. If the action is purely accidentally successful, the measure of V_0 is zero, and so is q, because V_0 will be an empty set.

Probability and goal-directedness

4.4. To indicate the theoretical relation between the degree of goal-directedness and the probability of G occurring, we may briefly say this.

If MU_0 is the measure of the set of all possible initial states of the system, and if we have no further knowledge of the system than that a fraction p of MU_0 results in G occurring and the rest do not, then the probability of G resulting from any one member of U_0 is p.

It follows that in a case in which we do not know which initial states result in G occurring and which do not, but do know that the system comprises an activity directed towards G with a degree of goal-directedness q, the probability of G must be greater than q.

Unsuccessful goal-directed behaviour

4.5. In section 4.2 it was assumed that the actual value of $\mathbf{u}_0$ is a member of the set S_0. The actual action therefore satisfies $F(\mathbf{a}_k, \mathbf{e}_k) = 0$, which we postulated to be a necessary condition for the success of the action. But this does not mean that the action must be successful. For a necessary condition is not the same as a sufficient condition. Other requirements may have to be fulfilled by the action. For example, a gun aiming at a target may in addition have to correct its elevation in accordance with the distance of the target. If it aims its sights correctly but fails to adjust them to the right distance, it satisfies our conditions for objectively goal-directed behaviour in the one respect but not in the other. In this sense therefore we can talk about objectively goal-directed, but none the less unsuccessful, behaviour.

This objective sense must, of course, be distinguished from the common subjective sense in which the actual result of a consciously purposed action is compared with its intended result. For reasons explained in section 1 this subjective concept is not admissible in the present context as it does not refer to objectively definable system properties.

Response functions

4.6. Biologists tend to think about animal behaviour in terms of 'stimuli' and 'responses', whereas engineers compare the 'input' of a servo-mechanism with its 'output'.

The notion of 'response' is not a simple one. We can equate the 'output' of a machine with the motor activities of a living organism, but not with its 'responses'. Suppose that in a laboratory experiment a cat on two occasions reacts to a given situation by pressing a lever. The biologist would then say that the response was the same. But, of course, the motor activities were not the same. The cat may have approached the lever from one side the first time and from the other the second time. What was the same on both occasions was the *goal* of the activities. We cannot escape this by simply saying that we call the response the same because the end result was the same. If on the second occasion the cat had panicked and accidentally tumbled against the lever, we would not have called the response the same.

To be precise we must say that on both occasions the actions were certain functions of the environmental variables and that these functions satisfied the conditions for goal-directedness we formulated in section 4.2 – and in respect of the same goal-event. In a precise analysis of the system properties the emphasis therefore must be on the functions linking action variables with environmental variables; we shall call these *response functions*. In sections 4.1 and 4.2, $A(\mathbf{u}_0)$ was a response function as here defined. But the term is not intended to refer only to those functions in which the argument is an initial state variable.

A *stimulus situation* may then be defined as the aggregate of the values of the independent environmental variables that enter as arguments into a response function. And, in particular, a *stimulus* may be defined as any binary variable that is defined on this aggregate. This would conform with the practice of regarding a stimulus as a factor that is either present or absent.

The input to the exteroceptors is a function of the environmental variables, and also therefore of the 'stimulus situation' as defined above.

The animal's motor reactions transform its environment and/or its relations to the environment. To that extent they generally result in a transformation of the sensory input, in fact in an *input transformation*. And the precise relation between a response function and the exteroceptor input is that the response function represents an *input transformation operator*.

The first lessons the young animal learns, as it 'experiments' with different response functions, are to associate its motor activities with the resulting input transformations. From these lessons it derives more or less fixed *transformation expectations* which determine much of its later behaviour.

Appendix: a quantitative example

4.7. The purpose of the present investigations is purely analytic and descriptive. The task in hand is to find precise ways of describing and analysing certain distinguishing characteristics of organic activities and vital organization; to find a set of basic concepts that enable us to deal with these characteristics in a way that shows their logical relation to the concepts of physics and engineering; and to diminish the philosophical and methodological confusions which they have so often engendered in the past. In that context it is immaterial whether quantitative and explicit formulae can be found for the mathematical functions that appear in the analysis. Unlike the engineer we are not concerned with predictions but with conceptual clarifications.

All the same it may be helpful at this stage to illustrate the analysis with a quantitative example – one that can be sufficiently simplified to enable simple quantitative formulae to be substituted for the functions, *F*, *A*, and *E* in section 4.2.

As an elementary example of goal-directed behaviour, take the case of a player kicking a ball which is passing within easy reach of his foot. For simplicity assume that the ball travels with constant speed and direction. Also assume that the actual kick is a

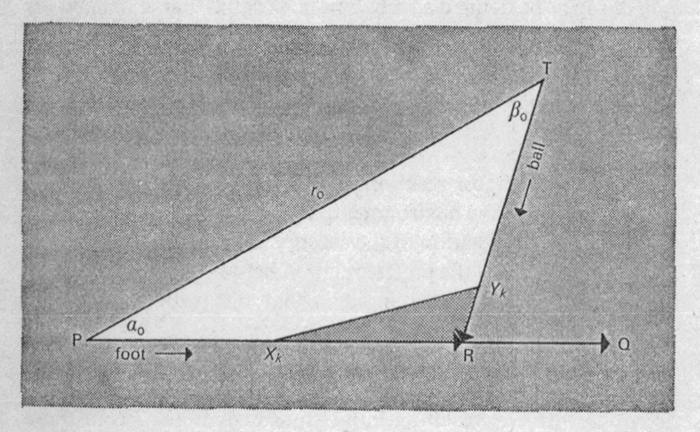

Figure 2

stereotyped action pattern, in the sense that once released the foot continues in a constant direction – and, we shall assume, also with a constant and given speed. Let the position of the ball and the foot be given in polar coordinates, using the rest-position of the foot as reference point (P in Figure 2) and an arbitrary line *PQ* as reference line. Moreover, let us assume that the objective goal of the action is to send the ball along *PQ*. The example therefore is of the goal-directed timing of a set action.

Let the initial time t_0 be taken as the instant that the ball appears in the player's field of vision, and let the position and direction of the ball at that instant be given by α_0, r_0 and β_0.

The only relevant decision the player has to make in this context is exactly at what point of the ball's progress he must start the kick. Let t_1 denote the point of time he chooses.

According to our analysis, whether the action is merely accidentally successful or whether it is a true instance of goal-directed behaviour depends on whether the player's response function results in the correct t_1 only for the actual case, or whether it is such that it would have produced the correct response also for a variety of alternative cases, i.e. alternative values of α_0, r_0, and β_0. And in the present case we may as well interpret 'variety' to mean the full range of values of these three variables consistent with our basic assumptions.

To find the conditions that his response function must satisfy in the second case, take any arbitrary point t_k during the kick. Let X_k and Y_k denote the position of the foot and ball respectively at that instant. We then have as a necessary condition for a successful kick

$$u \times RX_k - v \times RY_k = 0 \qquad 6$$

where u and v are velocities of the ball and foot respectively. This, therefore, is the form equation 1 (page 167) takes in the present example.

Transforming to t_0, we get:

$$RY_k = RT - ut_k = \frac{r_0 \sin \alpha_0}{\sin (\alpha_0 + \beta_0)} - ut_k$$

$$RX_k = RP - v(t_k - t_1) = \frac{r_0 \sin \beta_0}{\sin (\alpha_0 + \beta_0)} - vt_k - vt_1$$

as the formulae corresponding to equations 2 and 3 (page 168). Substituting these expressions in equation 6 gives:

$$\frac{ur_0 \sin \beta_0}{\sin (\alpha_0 + \beta_0)} - \frac{vr_0 \sin \alpha_0}{\sin (\alpha_0 + \beta_0)} - uvt_1 = 0 \qquad 7$$

This is the formulae corresponding to equation 4 (page 168) and so it gives us the conditions that the player's response function must satisfy if the action is to be goal-directed in the sense of making the attainment of the goal independent of variations of α_0, r_0, and β_0 despite the fact that each of these variables is relevant in deciding the course of events.

A practised player times his kick intuitively and no one can say what factors in the ball's movement the 'mechanism' of his 'intuition' actually evaluates in timing the kick. Whether it operates mainly by successive estimates of the ball's distance or goes by the angle the ball sweeps out in the player's field of vision, we do not know.

But the example shows that our concepts nevertheless enable us to make definite statements about his response function; viz. in this case that it must satisfy equation 7. With simple trigonometry, for instance, it could be shown that if the mechanism of his intuitive timing acted according to the formula

$$\sin \alpha_1 = r_0 \frac{\sin (\alpha_1 - \alpha_0)}{t_1}$$

it would satisfy the conditions (equation 7). Although the analysis does not uncover the response function, nor therefore the mechanism involved, it gives us some significant information about both.

5. Directive Correlation and Examples

Definition of directive correlation

5.1. The analysis of sections 3 and 4 has exposed the pattern of functional relations that is distinctive of goal-directed activities. It has shown what objective system properties they represent. A summary of the results was given in section 4.2.

It is now expedient to comprise these characteristic system properties in the definition of a single specific concept, which we shall call '*directive correlation*'. This will comprise all the relations that were brought together in section 4.2, but in a somewhat generalized form. And hereafter this specific and accurately defined concept will be available to replace our originally undefined concept of objective goal-directedness. It is generalized in two respects. In the first place, the division of all variables into action variables and environmental variables will be dropped as an unnecessary limitation. Secondly, it is now also expedient to dispense with the restriction of the function F to two independent arguments only.

If,

i. At a given time t_k it is a necessary condition for the subsequent occurrence of a certain event or situation G that two or more orthogonal variables $x_1, x_2, \ldots, x_n$ should satisfy

$$F(x_{1,k}, x_{2,k} \ldots, x_{n,k}) = 0.$$

ii. There exists an antecedent point of time t_0, a variable **y** and a set of functions $X_{1,k}, X_{2,k}, \ldots, X_{n,k}$, such that

$$x_{i,k} = X_{i,k}(\mathbf{y}_0) \qquad i = 1, 2, \ldots, n$$

where $\mathbf{y}_0$ is the value of **y** at t_0.

iii. There exists a set S_0 of values y_0 which has more than one member and is such that for all y_0 in S_0

$X_{i,k}(y_0)$ is bi-unique,
$i = 1, 2, \ldots, n$, and
$F[X_{1,k}(y_0), X_{2k}(y_{,0}), \ldots, X_{n,k}(y_0)] = 0$,

then $x_{1,k}\ x_{2,k}, \ldots, x_{n,k}$ will be said to be *directively correlated* in respect of the *goal G* and the *coenetic variable* y_0. The function F will be referred to as the *focal condition*. Unless stated otherwise, the set S_0 is assumed to contain the actual value of y_0.

The *degree* of the directive correlation may be defined exactly analogous to the degree of goal-directedness defined in section 4.3. And it follows from section 4.4 that the greater the degree of directive correlation the smaller is the element of chance-coincidence in the occurrence of G.

Again, if all functions are differentiable and y_0 is continuous in S_0, the directive correlation is characterized by

$$\sum_i \frac{\partial F}{\partial X_{i,k}} \frac{\partial X_{i,k}}{\partial y_0} = 0,$$

but not identically, for all values of y_0 in S_0.

All the variables in the above definition may be generalized variables. The only restriction is that they all refer to the state of the system at definite points of time. Their relation in time and their interdependence may be set out as in Figure 3. (Although we have here defined directive correlation in terms of a deterministic model of the system, parallel concepts can be defined on probabilistic models too. The uncertainty that commonly attends the behaviour of systems as complicated as the living organism can be taken account of in terms of the theory of *stochastic processes*. In the stochastic model of a system the state of the system at time t is defined in terms of the probability distributions of the alternative values that any physical variable of the system may assume at the time. In as much as in this conception the probability distributions at any time $t > t_0$ are uniquely determined by the probability distributions at time t_0 the system is deterministic in terms of these probability variables and concepts analogous to directive correlation can be defined

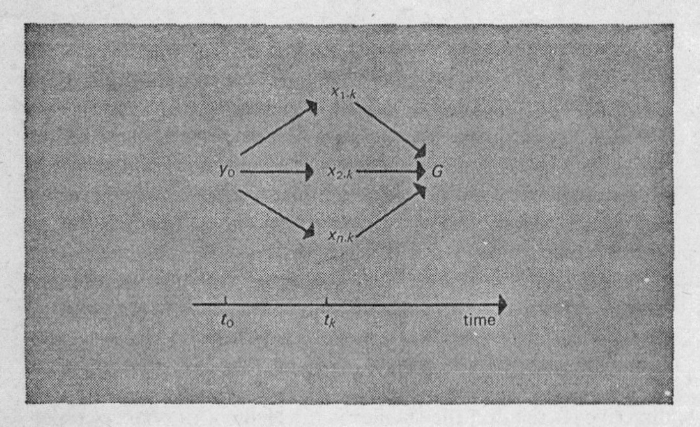

Figure 3

on them. Similarly, directive correlation may be defined in terms of the concepts of *information theory*, i.e. in terms of information transfer functions. But for the purposes of this paper the deterministic model we have used is perfectly adequate.)

The interval $t_k - t_0$ will be called the *back-reference period* of the directive correlation.

The black box and the football player

5.2. The directive correlation produced by the black box in section 3.2 is very simple. The necessary condition for a hit is that the line of fire should coincide with the direction of the red ball. Let these two variables be represented by the angles α and β respectively, and let t_k be the time of firing. The directive correlation is then between the variables α_k and β_k; and the focal condition is $\alpha_k - \beta_k = 0$.

As initial time t_0 may be taken any point of time that precedes

t_k by at least the operation lag of the mechanism. The coenetic variable is β_0. As the red ball is stationary we have $\beta_k = \beta_0$ and the response function of the mechanism must satisfy:

$$\alpha_k = \beta_0$$

The last two equations are the form that the set of equations given under ii, section 5.1 takes in the present case; and they clearly satisfy the conditions given under iii, section 5.1.

In the example of the player kicking a ball, section 4.4, the directive correlation is between the variables RX_K and RY_K. The focal condition is equation 6, page 173, and the coenetic variables are α_0, r_0, and β_0. The back-reference period is $t_K - t_0$.

The concept of adaptation

5.3. The concept of adaptation is one of the most general and important concepts that the biologist uses when dealing with the directive aspects of vital activities. But it has a number of different meanings which can lead to confusion if they are not carefully distinguished. When a chick pecks at a grain we may say that its movement is adapted to the position of the grain on the ground. We may also say that it has learnt to discriminate between grains and other small objects and that this learning was an adaptive process. And we may talk of the shape of the beak as a phylogenetic adaptation. In all these three cases we are referring to a specific form of directive correlation and it is only in terms of the details of this correlation that the essential differences between these three applications can be made clear. We shall find that the main difference lies in the length of the back-reference period (as defined in section 5.1); in other words, in the magnitude of *a variable that is never explicitly referred to in ordinary discourse and only emerges as the result of a precise analysis*. This shows the enormous importance of such a detailed analysis before any precise definition of a concept like adaptation is attempted.

In the first sense of the term, the concept of adaptation can be applied to both the examples detailed above; our black box, one can say, *adapted* its line of fire to the direction of the red ball, and the player *adapted* his kick to the path taken by the ball. In both cases the concept of adaptation has the full connotation of that of directive correlation. To say that the box adapted its line of fire to the direction of the red ball implies not only that the line of fire was appropriate in the actual case, but also that if the ball had been in a different position the line of fire would have become modified accordingly. And similarly with the player. Distinctive of the directive correlation of both these examples is the shortness of the back-reference period – a few seconds in the case of the black box, a fraction of a second in the case of the chick. Directive

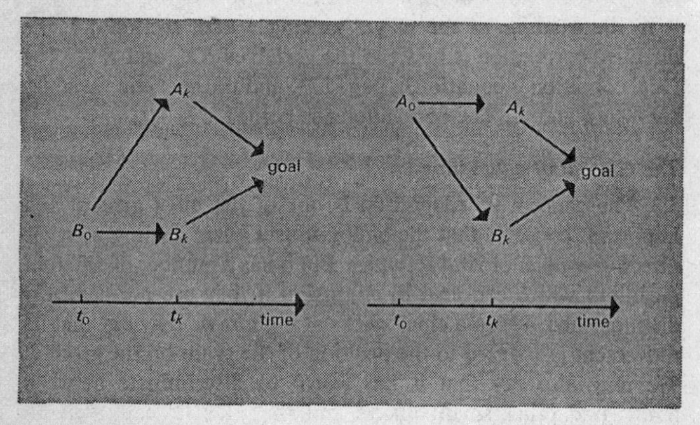

Figure 4a A adapted to B

Figure 4b B adapted to A

correlations with much longer back-reference periods will be encountered below, and this as has been said, is where the main distinction lies between the main types of adaptation encountered in nature. Adaptation is an *asymmetric relation*. '*A* is adapted to *B*' is not the same as '*B* is adapted to *A*'. The box adapted its line of fire to the position of the red ball' is quite a different proposition from 'the position of the red ball was adapted to the line of fire of the box'. This difference, again, can only be made explicit in terms of our concept of directive correlation. For it is simply a difference in the coenetic variable. In the first case β_0 is the coenetic variable. In the second case it would be α_0. In terms of our ideograms of directive correlation the difference between '*A* is adapted to *B*' and '*B* is adapted to *A*' may be represented as shown in Figure 5.

It is also illuminating to show the difference in terms of the 'lines of behaviour' defined in section 2.4. Suppose A and B are simple measurable variables and let the time variable t be taken as a third Cartesian coordinate. Let the focal condition of the

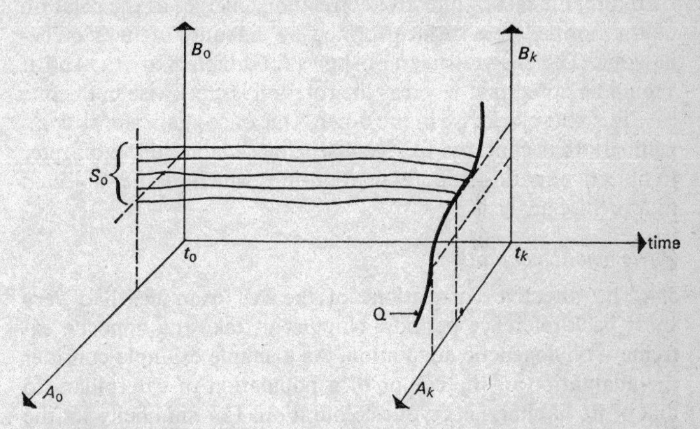

Figure 5a A adapted to B

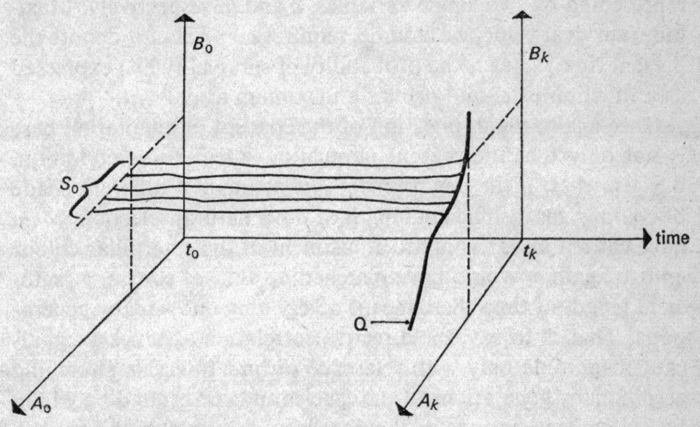

Figure 5b B adapted to A

directive correlation be $F(A_k, B_k) = 0$ and let this be represented by the curves Q in Figure 5a and b above. The first diagram represents 'A adapted to B'. The coenetic variable therefore is B_0, and each possible value of that variable is the origin of a line of behaviour. The fact that A_k and B_k are directively correlated now means that every line of behaviour that starts from a B_0 within the set S_0 must intersect the curve Q. In this representation, therefore, the fact of directive correlation emerges as the rotation of the normal of a surface formed by a family of lines of behaviour. The second diagram shows 'B adapted to A'. And it should be noted that whereas the rotation is clockwise in the one case it is anti-clockwise in the other. This once again shows compellingly that concepts like goal-seeking, adaptation, etc., refer to objective spatio-temporal relationships which can be subjected to a precise analysis.

Phylogenetic adaptation

5.4. The directive correlations of the last examples had very short back-reference periods. Now let us take the opposite extreme – phylogenetic adaptation. As a simple example consider the adaptation of the colour of a population of caterpillars to that of its habitat, i.e. cryptic colouration. For simplicity let the (modal) colour of the caterpillars and that of their habitat be represented by two single variables, c and h, respectively. Maximum survival value, we assume, requires $c = h$. Let t_k denote the present time; so at t_k the probability of survival (when expressed as a function of c_k and h_k) has a maximum at $c_k = h_k$.

Once again, the implication of the concept of adaptation here is not only that the present population satisfies this condition, but also that, *if* the environment *had* assumed a different shade of colour over a sufficient length of time, natural selection *would* have caused an appropriate adjustment in the caterpillar colour and so again c_k would have matched h_k. But, of course, a 'sufficient length of time' here means a very long one – many generations. That is to say, the directive correlation assumes a significant magnitude only with reference to time intervals stretching across many generations of the species and correspondingly long back-reference periods; t_0 having been chosen accordingly, we then have a directive correlation in which $c_k = h_k$ is the focal condition, and h_0 the coenetic variable. And the further back in time we take t_0, the greater will be the degree of the directive correlation, i.e. the measure of the set S_0.

That is no exception; *in many cases of directive correlation the degree of the correlation depends on the choice of* t_0. For instance, it would generally be true in our example of the football player. The shorter the back-reference period taken, the smaller would be the degree of directive correlation. In the case of the player, as t_0 approaches t_1 – and in the above case as t_0 approaches t_k – the degree of the directive correlation approaches zero.

It will be noted that the formal structure of the directive correlation in the caterpillar example is exactly homologous with that of the black box example. The only difference is in the length of the back-reference period required. So we may call the caterpillar case one of a *long-term* directive correlation, and the black box (or football player or chick pecking at grain) a case of *short-term* directive correlation. Our final examples will be *medium-term* ones. These characterize the ontogenetic adaptations, e.g. the processes of learning.

Learning

5.5. We are not here, of course, concerned with theories of learning, but merely with an accurate description of the directive element in the learning process. Speaking generally the directive element here consists of a directive correlation in which the appropriate back-reference period extends over the life span of the animal during which the learnt responses were acquired, e.g. the training period in a typical laboratory experiment.

Let a maze have N choice points. If arbitrary numbers are assigned to each of the possible choices at each choice point (e.g. 1 for turning left, 2 for turning right) the correct path through the maze may be represented by a vector **m** of N numerical components. An actual run through the maze can also be represented by a vector **r** of N components if some such stratagem is used as assigning 0 to all components after the first wrong turning. For instance, if there are five choice points and the correct run is LRRLR, **m** would be the vector (1, 2, 2, 1, 2); if the rat takes the wrong turning at the third point, **r** would be (1, 2, 1, 0, 0). Let the training time be measured not in standard units but in terms of the number of trials run. And let the rat take P trials to learn the maze.

Then, if we look at this instance of learning as a whole and not at the part-events involved in the process (section 5.9), the adaptive element in this instance is seen to be a directive correlation in which the correlated variables are $\mathbf{m}_k$ and $\mathbf{r}_k$, the focal condition is $\mathbf{m}_k = \mathbf{r}_k$, and the coenetic variable is $\mathbf{m}_0$, where t_0 must precede t_k by at least P trials. The assumption being, of course, that a maze of the same structure was used during the training period, which implies $\mathbf{m}_k = \mathbf{m}_0$. The set S_0, defining the degree of the correlation, here is the set of all (N^2) possible values of $\mathbf{m}$ in a maze of five choice points and two choices at each point.

Learning to discriminate

5.6. As another example of medium-term directive correlation we may take the case of a bird that has been taught to discriminate between N different geometrical shapes painted on the lids of a number of N similar food bowls, only one containing the food. Let x be the shape indicating the food and y be the bowl chosen, (to make these numerical variables, the shapes may be arbitrarily numbered). Again let the bird learn the discrimination after P trials. Let the time again be indexed by the number of trials and let t_k be the present time (trial). If the bird now chooses the correct bowl then the non-random element in this choice is due to the directive correlation established during the training period. And the details of this directive correlation are the following: the correlated variables are x_k and y_k, the focal condition is $x_k = y_k$, the coenetic variable is x_0, the training conditions are that x is constant (hence $x_k = x_0$), t_0 must be chosen to precede t_k by at least P trials, and S_0 is the set of the N positions (bowls) in which the food might have been placed.

Self-regulation and feedback

5.7. As an example of self-regulation, consider an imaginary black box with a radar scanner locking on a moving target and keeping the image of the target at the centre of a display screen. The case is analogous to visual fixation. Just taking a plan view, let α and β represent the direction of the object and radar aerial (or eye) respectively, and let dt denote the operation lag of the mechanism (including the period of transient response). Assume that this is a closed-loop control system – any deviation D of the image from the centre of the screen (or fovea) generates an error signal which the mechanism evaluates in producing the necessary corrections.

We can represent this case as a *continuous* directive correlation in which for any point of time t, α_t and β_t are directively correlated in respect of the goal $D = 0$, the focal condition being $\alpha_t = \beta_t$, and the coenetic variable being α_0; always provided that t_0 is

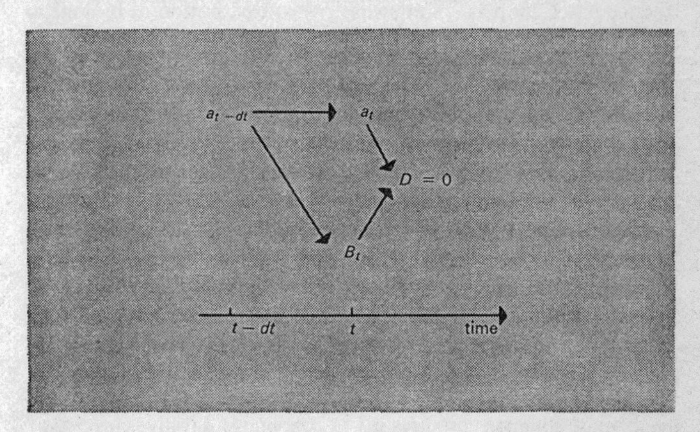

Figure 6

chosen to precede t by at least dt. As this set of conditions continues steadily throughout time, we may write $t - dt$ instead of t_0 to express the fact that we are here dealing with an invariant pattern of functional relations, a constant pattern that is independent of the time t. This continuous directive correlation may thus be represented by Figure 6.

Although the diagram contains an obvious loop, this is not, of course, the so-called 'feedback loop'. Feedback appears as a loop only in a diagram in which the time variable is implicit. 'Feedback' would here be taken to refer to the fact that the mechanism determines B_t in accordance with preceding error signals. But the error signal at any instant is a function of the values of a and B at that instant. And so, although we cannot point at this arrow or that in our diagram as representing the feedback, all the functional relations indicated in the diagram enter at one instant or another into the operation of the feedback mechanism. The concept of directive correlation is entirely compatible with that of the feedback loop, but it is more general and better suited to express what is most characteristic in biological situations and to clarify the overt characteristics of the system in a general theoretical context. As we shall see more clearly when we come to consider error-controlled servo-mechanisms (section 7.2) the notion of the feedback loop is a somewhat over-rated notion in biological contexts. Stripped of the engineering implications that are unwarranted in the biological applications it does no more than reaffirm what is already amply evident, viz. that the interaction of organism and environment is a reciprocal one.

Co-ordination

5.8. When a number of activities are co-ordinated – and it does not matter whether they are the part-activities of a single organism or whether they are the activities of separate individuals (as in a football team) – we can say this:

i. They have a common goal, G.
ii. In pursuing this goal each activity takes account of all the others.
iii. In most common instances it is a continuous directive correlation (see above).

In terms of directive correlation: if $x_1, x_2, \ldots, x_n$ represent the activities in question and t_k, t_0 have the usual meaning, the characteristic nature of the case may then be expressed by:

i. $x_{1,k}, x_{2,k}, \ldots, x_{n,k}$ are directively correlated in respect of the goal G.
ii. $x_{1,0}, x_{2,0}, \ldots, x_{n,0}$ are *all* coenetic variables in this directive correlation.
iii. t_0 may be taken as $t_k - dt$ (see above).

As an example of muscular co-ordination take the case of my hand moving in a straight line towards an object on the table. For simplicity let us assume that the shoulder remains stationary, and indeed let this be taken as a fixed reference point *Sh* (see Figure 7). The upper arm, lower arm, and hand may be represented by three vectors of constant length, **a**, **b**, and **c**, joined end to end.

If **p** is the position vector of the object, the distance **s** from fingertip to object is $\mathbf{s} = \mathbf{p} - (\mathbf{a} + \mathbf{b} + \mathbf{c})$. As we require that the hand moves in a straight line, we have as an example of one focal

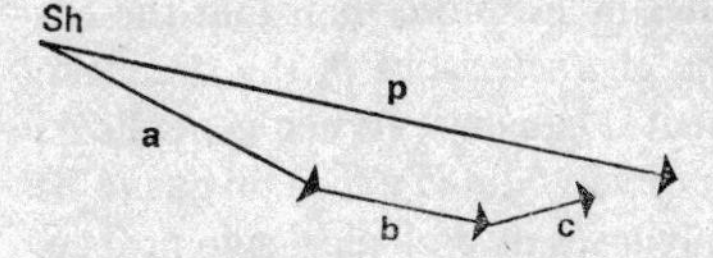

Figure 7

condition for the directive correlations involved that the vector product $\mathbf{s}_k \times \frac{d}{dt}(\mathbf{a}_k + \mathbf{b}_k + \mathbf{c}_k) = 0$. The coenetic variable is $\mathbf{s}_0$. Hence the functional relations required by the directive correlation are as in Figure 8.

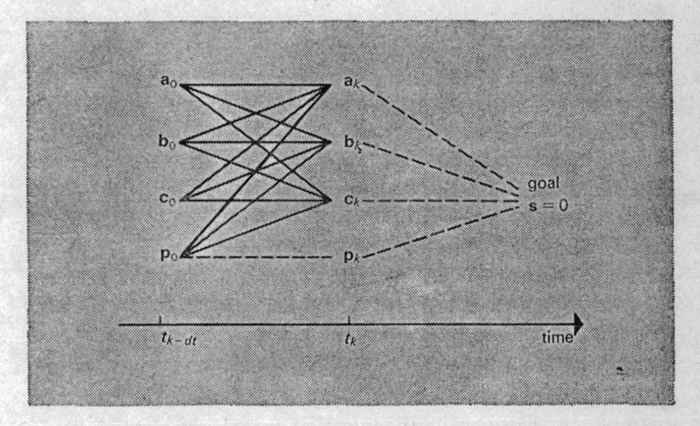

Figure 8

The functional relations shown in solid lines are the ones that must be established by the central nervous system, each requiring both afferent and efferent pathways. But in computing from this the minimum theoretical complexity of the network required it must be remembered that these are vector variables, the direction of every vector requiring the action of a set of several muscles.

Instinctive behaviour

5.9. Of the various concepts we have met in section 5, none is more frequently misinterpreted than that of adaptation. Adaptation, some textbooks tell us, means no more than that the environment is what it is and the characteristics of the organism are such that in that environment it can exist. When generalized to this degree, the concept acquires so diluted a meaning as to be virtually meaningless. It certainly loses its biological significance. What has been ignored, of course, in this interpretation is the element of directive correlation which we discussed in section 5.4.

Instinctive behaviour shows a high degree of phylogenetic adaptation, but that is not the most remarkable thing about it. Other forms of behaviour show that too. The real hallmark of instinctive behaviour is the comparative *absence of medium-term directive correlations.*

The difference between short-term, medium-term, and long-term directive correlations was stated in section 5.4. Every directive correlation has a precise meaning only in relation to a specific back-reference period which must be stated if confusion is to be avoided. Phylogenetic adaptation yields significant directive correlations only in relation to back-reference periods which extend over many generations of the species. This is 'long-term directive correlation'. In the 'medium-term' directive correlations the back-reference period can be taken within the lifetime of the individual (as in the case of learning). And as short-term directive correlations we denote the instant adaptations in which the back-reference period need not exceed a matter of seconds or minutes in order to yield significant degrees of correlation. Instinctive behaviour therefore shows high degrees of long-term directive correlations. It also usually involves various short-term directive correlations in the form of error-controlled and co-ordinated movements. What it lacks are the medium-term directive correlations, the elements of learning and ontogenetic adaptation. That is what we mean by the 'blindness' of instinctive behaviour patterns.

6. Organic Integration

The concept of integration

6.1. In section 5 we defined the mathematical concept of 'directive correlation' and illustrated how this concept enables us to express in precise terms the objective goal-directedness of organic activities and processes. It also enabled us to show at once the formal similarity of different patterns of adaptive behaviour and yet reveal their essential differences.

An important point to note is that in all cases the clarification of both the similarities and the differences compelled us to refer to variables which are never explicitly mentioned in the ordinary verbal description of the phenomena concerned (viz. the coenetic variable and back-reference period), and which only emerged as the result of an exact analysis. This shows not only how important it is to undertake conceptual analyses of this kind as an essential preliminary for making precise such concepts as adaptation, co-ordination, learning, instinct, etc., but also as a preliminary for unravelling the different kinds of goal-directedness that are usually present in the same biological situation. In the writer's mind there is not a shadow of doubt that it is impossible to arrive at a clear conception of the nature of organic order and the characteristic organization of living systems without an abstract conceptual analysis of the type intimated in these pages. Nor can we hope to make precise the concepts that are necessary for the scientist's thinking if he is to come to grips with what is most truly distinctive in the behaviour of living systems.

But although the goal-directedness of organic activities is the most fundamental property in which the distinctive organization of living systems reveals itself, this is not the whole story. Equally important is the way in which the different goal-directed activities of living organisms are mutually integrated – the way in which they combine to unite the system into an actively self-maintaining, self-regulating, and reproducing organic whole.

So far we have taken rather an atomic view of vital activities. We have concentrated on single actions or processes without considering their relations to others. And even in the case of a single action we have defined directive correlation merely in respect of discrete points of time. In fact, it may have worried the reader that directive correlation was defined only in terms of discrete points of time during the action. The concept of goal-directedness, he may say, applies to the action as a whole. And this is of course true.

In one respect there is no real difficulty here. We may call the whole flight of a bird a fast flight if at every or most points of time it had a high velocity. Similarly we may call the whole movement of the bird's beak directively correlated to the position of the grain on the ground if there was such a correlation at every or most stages of the action.

But when we have an integrated set of activities, as for example when I pick up the fork on the table and then proceed to eat with it, the situation is not so simple. The crucial point here is that we may look upon such a complex action in two equally legitimate ways. On the one hand, we may regard it as a *single* action with a single goal – eating my food. On the other hand, we may equally well regard it as a *sequence* of part-activities, each having its specific goal – moving my arm to bring the fork within reach of my hand, grasping the fork, lifting the food with the fork, and taking it to the mouth. And we would say that this was an *integrated* sequence of movements. *The concept of an integrated sequence of activities, therefore, stands for a relation between these activities which enables us to attribute an individual goal to each, and at the same time an ultimate goal to the whole sequence.* Our next step must be to find the exact nature of this relation.

The integration theorem

6.2. To this end we shall prove an important theorem which follows from the definition of directive correlation.

Theorem. If G_A is the goal-event of a directive correlation A and if the occurrence of G_A is a necessary condition for the occurrence of the goal-event G_B of a directive correlation B, then G_B is also a goal-event of A.

Proof. Let $F_A(x, y, z) = 0$ be the focal condition of the directive correlation A. Then, according to our definition, $F_A(x, y, z) = 0$ is a necessary condition for the occurrence of G_A. *Ex hypothesi* the occurrence of G_A is a necessary condition for the occurrence of G_B. It follows that $F_A(x, y, z) = 0$ is also a necessary condition for the occurrence of G_B. Hence in the sense of our definition, G_B stands in a relation of a goal-event not only to the directive correlation *B* but also to the directive correlation *A*.

In the sequel G_A and G_B will be called a *proximate* and *ulterior* goal of *A* respectively.

From this theorem it follows that, in the objective sense of goal-directedness, if the success of one goal-directed activity is a necessary condition for the occurrence and success of another, then the goal of the second may be called a goal of the first as well. To revert to the previous example, in as much as the movement of my arm towards the fork is a necessary condition for my fingers being able to grasp it, our definition of directive correlation entitles us to regard the proximate goal of the latter movement also as an ulterior goal of the former. A necessary condition for either movement is that the homeostatic mechanisms of the body efficiently regulate the physiological variables on which the action of the muscles depends – another example of the same type of integrating relationship. (The fact that the homeostatic regulation of the physiological variables does not serve just one specific type of muscular movement, but serves muscular movements in general – and that, therefore, we would not generally call a specific arm movement a 'goal' of the homeostasis – is irrelevant in this context. The formal relation between the two sets of directive correlations is the same.)

From this integral relationship between directive correlations, may be derived the precise concept of *hierarchies of directive correlations* integrated by a single ultimate goal. If the proximate goal G_C of an activity *C* is also the ulterior goal of a set of goal-directed activities B_1, B_2, . . ., in the sense defined, and if the proximate goal of each of these is in turn the ulterior goal of further sets of activities A_{11}, A_{12}, . . ., A_{21}, A_{22}, . . ., etc., we have a hierarchy of directive correlations united by the single ultimate goal G_C. This is the characteristic manner of integration of organic activities.

Sometimes the goal of a directive correlation at one level of the hierarchy is a correlated variable in a directive correlation at the next higher level. To illustrate again with the example of using a fork for eating: I grasped the fork because the food was solid.

If the food had been liquid I would have grasped the spoon. This is a directive correlation in which the goal of the grasping movement is correlated to the nature of the contents of the plate. This is a fairly typical situation. But it is no more than a special case of the general case discussed in this section.

The concept of hierarchy itself requires no analysis here. A full logical analysis may be found in J. H. Woodger's *Axiomatic Method in Biology* (Cambridge University Press, 1937).

The results of this section enable us to see in what precise sense we can say that in the living organism innumerable adaptive, regulative, and co-ordinative processes are combined to unite its component parts and their activities into an actively self-maintaining, self-regulating, and reproducing organic whole. Formally, the picture that has emerged is that of a physical system whose parts and part-activities are connected by complex hierarchies of directive correlations which have the necessities of survival and reproduction as ultimate goals at the apex of the hierarchy. The existence of this all-embracing system of directive correlations with its ultimate goal of self-preservation is what makes the organism into a living, biological unit.

The irreversible collapse of this system of directive correlations is the breakdown of its *animation*, is *death*. *Ageing* is the gradual shrinkage of the degrees of these directive correlations that precedes the final collapse.

Social integration

6.3. Complex patterns of directive correlations may exist not only between an organism and its environment, but also between members of an aggregate of individual organisms. And to the extent that these directive correlations form an integrated system of the kind outlined in the last section, they can impart a measure of organic integration and unity to such an aggregate. This may be merely a temporary state of affairs as when the activities of a football team are united during the game by a common purpose; or it may be a permanent condition, and if one of the ultimate goals of these directive correlation systems is to preserve the continued existence of the social aggregate for its own sake, the aggregate itself may come to function as a biological unit in a way formally not unlike that of a single organism. Typical examples could be taken from the social insects. The physical substrate of the interconnexions required may be much more tenuous than in a single organism and the channels of communication required may be merely punctuate in time, but from an abstract and analytical point of view the essential relations between the parts and whole are of the same general kind.

A living organism is a *physical* unit by virtue of the stable physical substrate that unite its parts into a single compact body; it is a biological unit by virtue of the hierarchies of integrated directive correlations of the type described in the last section. A social aggregate such as an insect colony is not a physical unit in the above sense, but it can still be a biological unit in the same sense as above. Provided there is communication between its members, the physical prerequisites are satisfied that enable the activities of one member to become functions of the activities of others. And so directive correlations are made possible into which the activities of the other members enter either as correlated or coenetic variables.

At the very general level of this discussion, therefore, the analysis of the abstract concept of social integration requires no basic concepts other than those which have already been set out in this paper. But this observation must be qualified in one particular respect, as we shall see in the next section.

Potential directive correlations

6.4. We have found that the goal-directedness of an action is an objective system property, and that to define it we must refer not only to what the system does under the given circumstances but also to what it would have done under certain alternative conditions. In this respect the property is not unique; many other physical properties can only be defined in the same general way.

We can now take this a step further. Suppose that in a given situation an animal is inactive but is in a state of *alertness*. It is hungry, let us say, and in a state of readiness so that as soon as a suitable prey appears it pounces on the victim – a thoroughly goal-directed action. It then follows from the general results of our analysis that we are entitled to regard this state of alertness also as an objective system property. For once again we can say that the animal *is* in such a state or condition that under such and such circumstances it *would* react in such and such a manner. And the precise manner of its reactions we would then specify in terms of the directive correlations involved in the action of pouncing on a prey – and more specifically in terms of the goal of these directive correlations. The state of alertness, therefore, is definable in terms of *potential directive correlations*, i.e. the directive correlations that would characterize its reactions if the environment were to change by the appearance of a prey. An animal in a state of alertness is not just in a state of potential activity, but of potential activity of a certain kind. And to define the kind we must refer to the directive correlations that the activity would show, and in particular to their goals.

The same analysis may be extended to the concept of *appetitive behaviour*. In appetitive behaviour we have a state of alertness coupled with a random exploration of the environment which lasts until the appearance of a prey, of food, water, or what have you, converts the behaviour into a discrete goal-directed action. Appetitive behaviour always has the appearance of being at once random and yet purposive. We can now see more clearly from what this double aspect derives. It is *random* because of the random nature of the goals of the *actual* directive correlations during the exploratory stage; it is *purposive* on account of the *potential* directive correlations during that stage, i.e. the directive correlations that become actual when the search ends as the prey comes into sight.

The same line of analysis may be used to give a precise meaning to the concept of *drive*. When an animal is subject to a certain drive, thirst let us say, it is objectively speaking in a state which is characterized by an extensive set of actual and potential directive correlations (more often the latter) which have the same ultimate goal – in this case the intake of water. The degree of these directive correlations denotes the strength of the drive, for it demarcates the variety of circumstances in which this goal over-rides all others. The weaker the drive the narrower this range of situations.

The concept of potential directive correlations is important in other respects too. For instance, in connexion with the idea of *social integration* that formed the subject of the last section.

In order to arrive at a concept of social integration that conforms with everyday usage, we must admit potential as well as actual directive correlations as the integrating factors of a social aggregate of organisms. There is no logical difficulty about this, because we have established both to be clearly definable objective system properties.

A social entity such as an insect colony or a family unit such as a pair of birds and their offspring, is not commonly looked upon as being a social unit merely during periods when its members are actually socially active. Even when its members are at rest we think of the family as persisting as a social entity. And the reason is, of course, that even when at rest its members are objectively so conditioned that a number of specific contingencies would at once evoke social behaviour of one kind or another. In other words, potential directive correlations are just as relevant to the concept of social entity as actual directive correlations, if not more so.

We may take the same detached and objective view of social units in human society. A family, a group of friends, a pair of lovers, a political party, a nation may be thought of in a detached and objective way as aggregates of individuals which are united by virtue of the goals of their potential more often than their actual behaviour. A patriot is not a patriot so much by virtue of what he does at every moment of his daily routine, but by virtue of what he would do in certain special contingencies, as in a case of a national emergency.

It may first seem strange to look upon these things in this objective and purely behaviourist way and to look upon, say, love and friendship as objective system properties. But that is precisely what the results of our analysis entitle us to do, and in doing this it takes us a step towards bridging the philosophical gap between mind and matter. For a great number of 'mental' states can be defined objectively in terms of the type of objectively goal-directed behaviour to which they would give rise in a variety of situations, i.e. in terms of potential directive correlations. It is along these lines that we may eventually succeed in bridging the gap between the language of psychology and the language of the physical sciences.

It is not within the scope of this paper to expand on this. But I may give one additional illustration. In section 6.2 it was said that death is the total breakdown of the integrated directive correlations that inform the living organism and, indeed, constitute what we would call its animation. Sometimes there is a partial breakdown of this kind, affecting merely the actual and potential directive correlations that exist between the current states of the animal on the one hand and the current states of the environment on the other (and indeed only the short-term ones), leaving intact, therefore, the internal directive correlations, the internal regulations, etc. This is then what we would call a state of *unconsciousness*. Now, consciousness may return without for a time overt behaviour returning as well. So what returns then are the potential directive correlations and not the actual ones. Formally the change is analogous to the change the system properties of an animal undergo when it passes from a state of non-alertness to a state of alertness. Except that we are now dealing with the totality of potential directive correlations. By following up this analytical line of approach we may eventually arrive at a scientific and objective definition of consciousness – long overdue as any neurologist will agree.

A mathematical definition of 'directive correlation' was given in section 5.1. It remains to state the formal difference between actual and potential directive correlations in the terms of that definition; it is a simple one. In an 'actual' directive correlation the actual value of y_0 is a member of the set S_0; in a 'potential' directive correlation it is not.

Learning by trial and error

6.5. Learning has been defined as 'that process which manifests itself by adaptive changes in individual behaviour as result of individual experience' (W. H. Thorpe, *Learning and Instinct in Animals*, Methuen, 2nd edn, 1963). Everything in this definition depends on the precise meaning of 'adaptive' in this context. This point was dealt with in sections 5.5 and 5.6. It amounts to specifying the pertinent type of directive correlation. In this section we shall look at a different question – the nature of the random element in trial and error learning.

Most learning involves a trial and error element. In this light the process of learning appears as one of a progressive substitution of co-ordinated and adapted responses for more or less random ones – and on a progressively increasing scale of complexity. From the point of view of a mathematical theory of learning mechanisms an important problem must be to pin down the exact nature of the random element. The same problem exists in the appetitive behaviour that an animal shows under the influence of a given drive.

To bring out the main point to be made here, consider this rough sketch of an imaginary but wholly typical learning process. The example is one of an animal that after birth first has to learn to walk and then to walk towards specific objects in its surroundings on which it may fix its attention.

At first even its individual leg movements are unsteady and random. Then they become increasingly organized. New response functions develop and with them new directive correlations: co-ordination of muscles to produce definite and coherent individual leg movements, co-ordination of leg movements to produce a directed movement of the body, co-ordination of body movement and eye movements to produce a steady advance towards the object. In fact, a hierarchy of directive correlations of the kind discussed in section 6.2 comes into existence. And it is a particular form of such a hierarchy – the goals of the directive correlations at one level appear as correlated variables in the directive correlations at the next higher level.

What are the random variables in this learning process? The hierarchy would obviously develop from the bottom upwards and it would seem that there is a twofold process. In the first instance the animal 'experiments' with different response functions at the lower end of the hierarchy, i.e. with individual leg movements. Gradually response functions of set forms become established. Whatever leg movements are produced from now onwards, they are internally co-ordinated ones. So from now on the 'experiments' consist of a random switching from one established response function to another, each giving a specific goal-directed leg movement. The new random variables, therefore, are the goals of the individual leg movements – the goals of their internal directive correlations. This stage lasts until these goals become mutually correlated in a directive correlation of a higher order and the established response functions become integrated into a more complex one. We have now arrived at the next level – co-ordinated leg movements producing directed body movement. And so the process goes on. So, as the process of learning advances, *the random element gradually moves up the organizational ladder*. And this continues right through to the advanced stages of trial-and-error learning. It never seems to drop back, there is never a reversion to chaos. A rat learning a maze by trial and error explores it systematically, i.e. by a random switching of goals at the highest level of directive correlation of which it is capable. Until again, when the maze has been learnt, a new and more complex response function has become established.

7 Brain-Like Mechanisms

Before saying a word about brain-like mechanisms, two different but widely held fallacies must be cleared up. One is that the goal-seeking behaviour of organic systems and servo-mechanisms is not fundamentally different from the equilibrium-seeking behaviour of inorganic systems. The other is the view that goal-seeking behaviour is co-extensive with error-controlled behaviour, i.e. with feedback control in the engineering sense. The first deprives us of any clues about the structure of the brain altogether. The second does have important structural implications but misleads us by presenting the brain as just a network of feedback loops.

Goal-seeking is not the same as equilibrium-seeking

7.1. Many biologists and cyberneticists tend to equate the characteristic directiveness of organic activities with the kind of directiveness that is shown by a physical system returning to a state of stable equilibrium. Although, of course, equilibrium-seeking processes of all kinds are essential ingredients of many organic activities, the characteristic goal-seeking quality of vital activities that has been our concern in this paper is of a different nature.

The dog pricks up his ears, jumps up, out of the door, down the stairs and welcomes his master in boisterous jumps – a whole sequence of goal-directed activities. Yet in what possible sense could each one of them be thought of as a physical system returning to a state of stable equilibrium (except perhaps in the completely trivial sense in which any state of activity is defined as a state of disequilibrium and any state in which the activity ceases as a state of equilibrium)?

Before going into the difference between goal-seeking and equilibrium-seeking in the strict physical sense let us look at the similarities. In both cases there is an end result which we may call a goal. It is reached from a variety of possible initial states. When a pendulum returns to its equilibrium position we might in a very broad sense call that a goal-event. And in both cases the occurrence of the goal-event presupposes that certain variables of the system satisfy certain specific conditions. In the case of directive correlation we called those the 'focal conditions'. In the case of the pendulum it is a necessary condition for a return to equilibrium that the force of acceleration is a certain negative function of the displacement of the pendulum. So much for the similarities. But now the crucial difference. In the case of the directive correlations the variables concerned in the focal conditions are orthogonal variables. (See sections 2.2, 4.1, and 4.2.) But in the equilibrium case the corresponding variables are not orthogonal. Thus in the case of the pendulum the force or acceleration is dependent on the displacement. Arbitrary combinations of these variables are not possible initial states of the system (section 2.2). This is the essential difference between goal-seeking in the biologically significant sense, and equilibrium-seeking in the technical sense of classical physics.

This is not just a purely formal difference. It goes deeper than that. There is much to be said for the idea that the *freedom of choice* we attribute to the behaviour of the higher forms of life amounts to no more than an implicit recognition that we are dealing here with complex physical systems whose overt action variables are orthogonal to the variables of their environment, i.e. that arbitrary combinations of both sets of variables are conceivable as possible initial states of the combined organism environment system. And the higher the animal on the scale of life, and the greater, therefore, the variety of actions we can see it perform in apparently similar circumstances, the stronger this idea establishes itself. In fact, it is interesting to speculate whether our idea of the *freedom of the will* is perhaps no more than just an expression of the fact that we look upon ourselves in this way, i.e. as a system in which arbitrary combinations of action – and environmental variables are possible initial states of any chosen time-slice. If there is any truth in this suggestion, then of course the whole apparent antithesis between freedom of the will and causal determinism would melt away.

Directive correlation not co-extensive with feedback control

7.2. Error-controlled devices, such as servo-mechanisms with feedback, can produce goal-directed behaviour in the full sense of our definition of directive correlation, provided they are seen in the broad context in which they are applied.

Consider a ship automatically held on its course by a servo-mechanism. The goal of the directive correlation then is that the ship sails along a certain course. The focal condition is that the output signal O (representing the ship's actual direction as registered by a gyro) equals the input or command signal I (which represents the desired direction and is derived from the setting of the controls on the bridge). This the servo-mechanism achieves except that, for instance, a sudden change in the command I may be followed by a short period of transient error before once again $O = I$. So for any arbitrary points of time t_k and t_0 the conditions of a directive correlation between O_k and I_k are satisfied, provided $t_k - t_0$ exceeds the period of transient response. The coenetic variable is I_0. The conditions of orthogonality are also satisfied, as arbitrary combinations of I and O are possible initial states of the system.

But the evident fact that error-controlled devices can yield directive correlations, and the further fact that obviously many individual instances of animal behaviour and vital activities superficially appear to be of a similar error-controlled type, does not justify the conclusion that all directive correlations found in animal behaviour, at any rate all short-term ones, are based on similar devices. There are many cases for which this is manifestly untrue – for instance, all actions involving single choices. I approach the door of my house and enter. Here the approach towards the door might possibly be interpreted as an error-controlled movement in which visual and proprioceptive impulses provide the basis of an error computation which is then used in determining the corrective outputs. But it does not apply to my choice of this door as distinct from the other doors of the street. And yet that is a directive correlation too. And even if we just look at the movements involved in approaching the chosen door, all we can say strictly speaking on the observed facts of the case is that these movements are *error eliminating* – not that they are *error controlled* in the strict sense of the term in which this implies a mechanism based on the initial computation and setting up of an error signal reflecting the magnitude and direction of the discrepancy between the actual and desired state of the system. The final output of the central nervous system reflects this discrepancy, of course. But we have no evidence that explicit error signals are set up at any of the intermediate stages in the processing of the sensory inputs in this particular context. In fact the plasticity required of the nervous system in adaptive activities of this particular kind argues against the employment of mechanisms requiring such error computations. The familiar 'comparators' of the servo-engineer imply functional rigidities that are difficult to reconcile with the degree of plasticity required in the central nervous system. And it is, of course, well known that the only evidence we have of a servo-mechanism in the strict sense in the nervous system involves merely the purely mechanical action of the stretch reflex.

It cannot be stressed enough that the indiscriminate application of engineering concepts to biological situations is fraught with danger. Only extreme caution and careful analysis can save us from the many possible pitfalls.

To sum up, directive correlation is a wider concept than error control and, though less easy to visualize when we look for 'mechanisms', it is biologically the more appropriate concept.

The mathematically minded reader may at first be puzzled by the fact that a servo-mechanism produces directive correlation and a pendulum does not, despite the fact that the behaviour of both can be described by very similar dynamical equations. The reason is that the dynamical equations relate neither to the context in which the appliance is used (i.e. to the formulation of the goal) nor to the question of orthogonality.

'Brain-like' mechanisms

7.3. Automata of all sorts are being designed today with the express purpose of mimicking the behaviour of living systems, particularly that of the brain. But mimicry of this kind can be a very superficial thing, and its importance for the biological sciences is easily over-rated if we gloss over the question of how close an analogy to living systems they really achieve.

A generation or so ago the public was entertained by fanciful robots in stylized human form which would walk and talk and flash their neon eyes. No one today would say that they had achieved any real significance in illuminating biological problems.

But how truly brain-like are our modern devices? Will their biological significance seem any greater to the next generation?

None of this can be decided until we achieve a much clearer conception of how the most generally distinctive functions of the brain are to be described in objective terms. Because of that we are still very much in the dark; and in the writer's opinion the only way to achieve progress in this respect is by an abstract and accurate analysis of the kind that has been outlined in this paper.

The fundamental functions of the brain throughout the animal kingdom are those of adapting, regulating, co-ordinating, and integrating the animal's activities. That is to say, they are functions which can only be defined accurately in terms of directive correlations and hierarchies thereof.

Once we have analysed and defined them in these terms, the mathematical definition of directive correlation yields equations which can serve as *criteria* for brain-like mechanisms – equations which can be used to assess how close an analogy to the living brain our artefacts achieve. The equations derived in this way will certainly not be sufficient to give us detailed design formulae; but they will be important design *directives* and give us a clearer idea of what to aim for.

Such an approach should also kill once and for all one of the major fallacies that are still current today, viz. the presumption that because the rules of mathematics and logic have found a physical incarnation in the activities of electronic computers, these machines can *ipso facto* be thought of as truly brain-like, or that their organization can throw any light on the functional organization of the brain. They are extensions of the known brain, not analogues.

Physiologically, the human brain differs in no radical respects from the brain of the nearest primates. It is not even very much larger, containing only about 20 per cent more neurons. And, speaking objectively, its primary functions have remained exactly the same, i.e. to generate adaptive responses to the problem situations with which the environment confronts the individual. What is new in the human situation is that with the arrival of speech and communication these problem situations are often *communicated* ones calling in turn for *communicative* responses. That is to say, the problem situation is presented in terms of the spoken or written signs of the human language and calls for response made again in terms of such signs. The correct responses therefore require a double adaptation – an adaptation to the rules that govern the established usage of the signs as well as an adaptation to the situations signified. Again speaking objectively, intelligence is an attribute relating to the brain's ability to cope with the adaptiveness required in both respects. All this means an immense increase in the complexity of the focal conditions of the directive correlations involved. But the complexity of a focal condition does not necessarily mean a corresponding complexity in the mechanisms that establish the directive correlations concerned. To take an example from phylogenetic adaptation: the mechanism of inheritance and natural selection is relatively simple compared with the complex end results it can engender.

Seen in the light of these objective criteria we are compelled to the conclusion that true machine intelligence is still a long way off.

Summary and Conclusions

1. The distinctive organization of living systems manifests itself in the goal-directedness of their activities (section 1).

2. This goal-directedness is an objective system property which can be expressed in terms of mathematical relations between physical variables – in fact a physical property (sections 3 and 4).

3. The mathematical definition of 'directive correlation' (section 5.1) illustrates one way in which this system property may be given a precise definition.

4. Goal-seeking is not the same as equilibrium-seeking (section 7.1) nor is it co-extensive with feedback control (section 7.2).

5. A mathematical definition of directive correlation enables many biological key-concepts to be made precise. On this foundation a bridge may ultimately be built between the language of biology (and psychology) and that of the physical sciences (section 5).

6. The main examples discussed were the concepts of adaptation, regulation, co-ordination, learning, instinct, and drive (sections 5 and 6).

7. The fact that these definitions required an explicit reference to 'hidden' variables such as the 'coenetic variable' and the 'back-reference period' demonstrated that these concepts had inherent ambiguities which only a precise analysis can eliminate.

8. The nature of the random element in trial-and-error learning was discussed in section 6.5.

9. Organic integration was explained in terms of hierarchies of directive correlations (section 6.1).

10. The logical structure of these was established by the integration theorem stated in section 6.2.

11. The objective substrate of social integration was outlined in sections 6.3 and 6.4.

12. Among the wider concepts that appeared in a new light were those of life and death (section 6.1), consciousness (section 6.4), and freedom of the will (section 7.1).

13. The significance for the neurologist of the various 'brain-like' mechanisms *that are being developed in our time* cannot be assessed until we have *clearer formulations of the most generally distinctive* functions of the brain. Here the concept of directive correlation, or the development of similar concepts, would seem to be an indispensable prerequisite (section 7.3).

14. If agreement can be reached on this point, the mathematical definition of directive correlation will then yield equations which can serve as criteria for brain-like mechanisms, and as directions for their future development (section 7.3).

Part Three The Environment of a System

Even among theorists purportedly dealing with open systems there has been a reluctance to come to grips with the problems of characterizing the environments within which systems must adapt to survive or duplicate themselves. These theorists have sought to cope with the problem by simply adding some exchange equation to the equations defining the inner system (see von Bertalanffy, Reading 4, page 70). This lands one with insuperable difficulties that seem to focus upon problems of defining the boundary conditions. Just how slippery these problems are has been spelt out by A. Angyal (*Foundations for a Science of Personality*, Harvard University Press, 1941, pp. 88–99) and can be easily inferred from efforts made by others to solve them.

Sommerhoff's analysis (Reading 8) was one that has shown that the boundary conditions or exchange processes are as much subject to the heteronomous processes in the environment as they are to processes within the system. Thus, open systems analysis cannot hope to stop with a specification of an exchange equation; it can hope to approach adequacy only when there is some characterization of the environment.

The reluctance to tackle environmental analysis appears to have arisen from the forbidding nature of two problems – (a) the sheer complexity of most environments and (b) the incommensurateness of the many heterogeneous processes that make up the system and its environment (e.g. psychological, economic, technical, meteorological).

Sommerhoff had, as far back as 1950, provided a perfectly satisfactory answer to the second problem. The first problem should not have been taken so literally. Not all aspects of an environment are equally relevant to any particular system or class of systems. In economics, military theory, ecology, psychology, to name but a few, there has been a long and successful tradition of identifying and characterizing global aspects of environment and relating these to adaptation. The readings in this section show something of the progress towards a general solution.

9 M. P. Schützenberger

A Tentative Classification of Goal-seeking Behaviours

M. P. Schützenberger, 'A tentative classification of goal-seeking behaviours', *Journal of Mental Science*, vol. 100 (1954), pp. 97–102.

Goal-seeking behaviour, once a great mystery, is now beginning to be understood. In its simplest forms it is, in fact, understood today almost completely. Thus the theory of the simple regulator, such as the thermostat, not only includes an extensive repertoire of techniques, but the elementary principles, of the necessity for negative feedback for instance, are becoming scientific commonplaces. In getting to know, however, about these simple systems and their principles, we should not make the mistake of thinking that there is nothing more to be learned. On the contrary, in real life many an important goal is to be achieved only through some quite complex pattern of behaviour, a pattern for which the simple concept of 'negative feedback' is quite inadequate. It is of these more complex patterns that I wish to speak today.

One way of studying the subject is by way of actual experiment; but I shall make little reference to actual experiments today. The fact is that before such experimentation can be undertaken with any usefulness there must be a preliminary period of study and thought. Before we can experiment we must be clear about what questions we want to ask, and why these questions are significant, and what are to be the interpretations of the experiment's various possible outcomes. Before we can usefully start experimenting, in other words, we must have a well-developed theory. Such a theory must inevitably, if it is to be precise, be mathematical; but I hope to show in this paper that what is necessary, at least at first, is the logic and precision of mathematical thought rather than its more advanced techniques. If then we are to explore the properties of the more complex forms of goal-seeking behaviour we must first construct some suitable mathematical models.

The use of mathematical models in the study of animal and

human behaviour goes back to the early nineteenth century, but those early attempts have little in common with the contemporary researches of such scientists as von Neumann, Wiener, and, here among us, Dr Ashby. The last author has particularly stressed the similarity between the activities of some types of mechanism and that of the brain; and this similarity must be the excuse for a mathematician such as myself venturing into this controversial field. My aim will be to show how certain of the more complex goal-seeking behaviours, seen in both machines and men, can be described, and the principles made clear, by a uniform mathematical framework of ideas.

Basic Concepts

In order to make the ideas as clear as possible I shall start with a very simple example. Let us suppose a man is on the top of a hill and that he wishes to get to a house in the valley; let us assume that the 'goal' is his arrival there in *the shortest possible time.* Between him and the house are many causes of delay: boulders. marshes, escarpments, and so on. Travelling in a bee-line is out of the question. Let us consider his possible modes of behaviour.

An exhaustive, and final, solution of the problem would be given by taking a map of the district, dividing it into small areas, finding the time taken to cross each area individually, joining the areas into all possible chains between the top of the hill and the house, and then finding which chain gives the smallest total time for the journey. The path so selected is absolutely the best and has been selected by what I shall call the 'strategy' of the problem. (I shall use the words 'strategy' and 'tactic' throughout this paper in the particular senses defined, and with no correspondence to other usages, though in some cases our 'strategy' will correspond to von Neumann's 'Minimax strategy'.)

Usually, of course, the traveller would not use so elaborate a method. A common method would be to make the selection in stages. He would first select a point about a hundred feet down to which he could get rapidly; then, arrived there, he would select another point a hundred feet lower still to which a rapid descent was possible, and so on till he reached the house. This method I shall call a simple *tactic*, as contrasted with the previous *strategy*. The tactic differs from the former in that the tactic does not take into account the whole of the situation, but proceeds according to a criterion of optimality that is applied locally, stage by stage.

This example, of the traveller on the hill, is really more general and widely applicable than might at first appear. We can in fact replace the hillside by an abstract space and the expected elementary times t_i (on the elementary areas) by some function $f(P)$ of the points P of the space. At this level of abstraction the problem is then to find a curve C between the initial and terminal points, such that the integral

$$\int_C f(P)dP$$

is a minimum.

To see more clearly how the concept can be applied when the problem is not geometrical, let us consider an example of quite a different type. Suppose a specialist in vocational guidance has to allot n given persons to n given jobs, when he has already tested them and has made predictions (n^2 in number) of the 'suitability' (on some scale) of each person for each job. His 'goal' is such an assignment as will maximize the suitabilities *over the whole set.* Thus, it might happen that three people A, B, and C had the suitabilities shown in the table for jobs I, II and III.

	I	II	III
A	9	10	4
B	5	8	4
C	3	2	1

An 'assignment' (of people to jobs) is determined when we have selected three entries, no two being in the same row and no two in the same column.

Here the strategy would consist of computing the totals corresponding to each of the $n!$ possible assignments. In this example there are $1 \times 2 \times 3$, i.e. six possible assignments; trials soon show that the best is that which gives A the job I, B the job II, and C the job III. This scores 18, and no other assignment scores more.

If n is large, the computations become prohibitively heavy; a possible tactic is then to proceed by first picking out the largest entry in the table, then cancelling out that person and that job, then selecting the next largest in the remaining $n - 1$ persons and jobs, and so on to the completion of the assignment. If it is applied to the table, then the score of 10 first determines that *A* is to be given job II, then the 5 gives *B* the job I and then *C* gets job III.

This tactic scores 16, little less than the maximum of 18. We see therefore that it may be possible to achieve fairly easily a score that is only slightly below the best when the best itself is quite impossible of achievement (as would have been the case in this example had *n* been thirty instead of three).

The same principle occurs in chess-playing. If, towards the end of a game, player *A* can see how to beat *B* in spite of all defences, then *A* will be following what I have called a strategy. Often, however, this perfect way to the goal cannot be perceived; then *A* looks for some move that will achieve a temporary or local advantage, such as capturing *B*'s queen, or covering the largest area of the board, or promoting a pawn, etc. Such a move is a tactic.

Strategy and Tactic

Our next step is to see more clearly what is the relation between these two – to show them as derivatives of a single concept. Let us go back to the man on the hill.

Let us assume that by now he has discovered everything about the hillside, so that he now knows the true minimal time T_i necessary for going from every point P_i of the hillside to the house. This information he has conveniently schematized on his map by drawing a series of *isochronic curves* such that each curve joins all those places that are situated at the same (minimal) time from the house. Thus, the curve marked '5 minutes' would run through all those places from which the house can be reached in 5 minutes but not less.

Once the map has been prepared, a simple reasoning, borrowed from the calculus of variations, shows that, given any starting point, *the optimal path to the house is one that cuts every isochronic curve at right angles.* (Technically we may express this by saying that the optimal paths are geodesics in a variation problem and that the isochronic curves are the 'transversal' of the problem.)

From the practical point of view, the introduction of the isochronic curves is not a mere artifice. In some problems a computation of the solution may be obtained easily if the path is computed backwards from the goal. Sometimes, though more rarely, it happens that computation of the isochronic curves is actually the quickest way to the solution – the most economical in the number of arithmetical operations. Thus, in the problem of assigning jobs we saw that $n!$ sums involving $(n - 1)\,n!$ additions had to be performed. If, however, we follow a method derived from the considerations just given, the number will not exceed $n(2^{n-1} - 1)$, which is far fewer; for instance, if n is 12, the number of elementary additions falls from 5,269,017,600 to 24,564. Similar reasoning has been used by Hoffman to design codes of minimal redundancy.

The introduction of isochronic curves enables us now to see more clearly the relation between a strategy and a tactic. For if a point follows a path that is everywhere at right angles to the isochronic curves then it is also moving in such a way as to make maximal the instantaneous decrease of the remaining time – it is moving optimally according to the *local* conditions. In other words, if our man behaved as if acted on by a field of force deriving from a potential (which had the isochronic curves as equi-potential lines), then his path would be identical with that determined by the optimal strategy. It is now clear that once the isochronic curves are given, i.e. *once the map's projection has been changed to that of the really optimal metric, the distinction between strategy and tactic disappears.*

Classification of Imperfect Tactics

Having considered the optimal path, let us turn next to consider the case of the path that is grossly non-optimal, to that, say, of a stone rolling down towards the valley under the action of gravity. According to the laws of physics, the stone, at each moment, is falling in such a direction as takes it the longest distance down in the shortest time, when only immediately local conditions are considered. Thus the difference between the behaviours (or paths) of man and stone is simply a difference between the fields which direct them; the man's field is truly optimal, for it is based on the isochronic curves, and the stone's field is that of its crudest approximation – the Newtonian field of gravitation.

In addition to these two fields are others; what can we say *a priori* of their properties and of their values as tactics?

First they may be classified according to what I shall call their 'span of foresight'; thus, if the man coming down the hill plans each next move according to the details of the next hundred feet he will do better than if he were to plan only over the next ten feet. The spans of foresight are here a hundred feet and ten feet respectively.

Should the span of foresight be equal to the distance from the goal, then obviously the tactic and the strategy become identical.

The second characteristic that is to be considered is the behaviour's 'flexibility'. Suppose the span of foresight is '100 feet below'; once the man has covered half this distance he may discover that his provisional goal was not the best, and that he should now take a different path for a different goal, again at one unit of foresight ahead. Clearly, in the strategy of the traveller with complete foresight the concept of flexibility plays no part, and neither does it at the other extreme – the case of the rolling stone (for the latter's steps are assumed to be infinitesimal, so that no smaller step is possible). It is in the intermediate degrees that the concept becomes important.

After these preliminaries, it will be seen that any goal-seeking behaviour may be classified on a scale of strategies and tactics, each one depending on a general function representing some measure of the distance between the present position and the goal. Thus, for the stone, the function depends on the present altitude above the bottom of the valley. Had the stone been a piece of iron and the goal a big magnet, the function would have depended on the object's present distance from the magnet. Other models would have led to other functions; so we have many tactics, each depending on a single function. The complete specification, and full classification, are given when to this function we add the span of foresight and the degree of flexibility, the latter being conveniently measured by the minimal time at which the provisional goal may be replaced by another.

The Stochastic Environment

What we have done so far is to show that the 'strategy' is simply one of the tactics: it is that extreme tactic based on the best function as given by the isochronal curves. It is now instructive to show conversely (so close is the relation between them) that any tactic may be viewed as some sort of strategy. This is so if the time t_i taken in crossing each elementary area is not permanently constant, as we have assumed so far, but depends on other factors of which we know only the probability of their values. Thus one of the areas might be a marsh that is easy to cross when the weather has been fine but difficult to cross after rain; if the traveller does not know what has happened prior to a *particular* trip, he can give only a probability to any particular 'duration of crossing'. Again, another area might be a lake, to be crossed only by a boat which may or may not be available, and so on. More generally we can assume that the elementary times t_i are not given fixed values but are given fluctuating values, depending on some random or 'stochastic' process.

If now we apply the theory of inductive behaviour as defined by Wald on the Ville–von Neumann principles, we find that *the optimal strategy is just the simple tactic of attempting to do one's best on a purely local basis.*

To illustrate this thesis let us consider some well-known examples. First consider that of the dog that wants to run to his master, who is himself walking steadily in a definite direction. If the dog is something of a computer it will perform an integration and will go directly to the place at which their relative velocities will enable them to meet. If the dog is not so clever it will continuously run directly towards where its master is at that moment. This second tactic is, of course, the simpler of the two, but is inferior if the master is moving uniformly. If, however, the master is making steps backward and forward in a totally random way, as if he were undergoing a Brownian movement, then this second tactic can be shown, by mathematical proof, to be actually *the best one possible*. The tactic has become the strategy.

Here is a second example. A number of lengths of telephone wire must be joined end to end to complete a long-distance line. Each length imposes some small but characteristic distortion a_i

on the message. The overall distortion is the sum of the individual distortions, but by reversing a length before it is joined the distortion can be added positively or negatively. How should the lengths be joined if the total distortion, $\Sigma \pm a_i$, is to be a minimum?

If a hundred lengths are to be connected, the absolute optimum can hardly be achieved, for the number of sums that would have to be calculated (if all possibilities are to be explored) is prohibitively large. What is the engineer to do? In practice, he adds the lengths one by one, at each new addition adding the wire this way or that so as to make the sum as small as possible at that particular stage. This tactic has been found to give quite satisfactory results, its success, perhaps, depending on the fact that this tactic really is the best possible if the wires are provided, and have to be connected, one by one.

Other examples of this thesis – that the optimal strategy often consists of doing one's best on a local basis – is also used in Fano's method for finding an optimal code, and in Gavrilov's method for building up a switching system that will represent a given logical function at minimal cost.

Conclusion

In this paper I have attempted to show something of what is implied by 'goal-seeking' when the whole situation is more complex than that occurring in, say, a simple thermostat. It is clear that the further study of these situations will have to be made mathematically. It is also clear that much more development will be necessary on the purely mathematical side, for much of the necessary mathematics will have to be developed specially. To the mathematician many of the problems raised are new, and will need new methods.

In particular we need to know more about how much efficiency is lost when the span of foresight and the degree of flexibility are not optimal. The problem is not made easier by the fact that often the parameters do not enter the problem with random values, so the appropriate theorems will have to be stated in a somewhat unusual form.

The purpose of this paper, however, was not to enter into these technicalities but to show in a general way how the introduction of mathematics into this branch of psychology might itself be a worth-while tactic.

Summary

When 'goal-seeking' behaviour is considered in situations of more than the most elementary type, the problems that arise are related to those of strategies and tactics.

I have attempted to show that clear-cut principles are involved, capable of mathematical treatment.

It appears likely that among the factors of special importance are those of 'span of foresight' and 'degree of flexibility'.

The case has also been considered in which the organism faces an environment that can be characterized only in terms of probability.

References

ASHBY, W. R. (1952), *Design for a Brain*, Chapman and Hall [2nd edn, 1960, Wiley].

GAVRILOV, A. I. (1949), *Theoria releino Kontaknykh Skhem*, Moscow, p. 185.

VON NEUMANN, J., and MORGENSTERN, O. (1947), *Theory of Games and Economic Behavior*, Princeton University Press, 2nd edn [rev. edn 1955].

WALD, A. (1947), *Sequential Analysis*, Wiley.

WIENER, N. (1948), *Cybernetics*, M.I.T. Press [2nd edn 1961].

10 H. A. Simon

Rational Choice and the Structure of the Environment

H. A. Simon, 'Rational choice and the structure of the environment', *Psychological Review*, vol. 63 (1956), pp. 129–38.

A growing interest in decision making in psychology is evidenced by the recent publication of Edwards' review article (1) and the Santa Monica Conference volume, (7). In this work, much attention has been focused on the characterization of *rational* choice, and because the latter topic has been a central concern in economics, the theory of decision making has become a natural meeting ground for psychological and economic theory.

A comparative examination of the models of adaptive behavior employed in psychology (e.g. learning theories), and of the models of rational behavior employed in economics, shows that in almost all respects the latter postulate a much greater complexity in the choice mechanisms, and a much larger capacity in the organism for obtaining information and performing computations, than do the former. Moreover, in the limited range of situations where the predictions of the two theories have been compared (see 7, ch. 9, 10 and 18), the learning theories appear to account for the observed behavior rather better than do the theories of rational behavior.

Both from these scanty data and from an examination of the postulates of the economic models it appears probable that, however adaptive the behavior of organisms in learning and choice situations, this adaptiveness falls far short of the ideal of 'maximizing' postulated in economic theory. Evidently, organisms adapt well enough to 'satisfice'; they do not, in general, 'optimize'.

If this is the case, a great deal can be learned about rational decision making by taking into account, at the outset, the limitations upon the capacities and complexity of the organism, and by taking account of the fact that the environments to which it must adapt possess properties that permit further simplification of its choice mechanisms. It may be useful, therefore, to ask: how simple a set of choice mechanisms can we postulate and still obtain the gross features of observed adaptive choice behavior?

In a previous paper (6) I have put forth some suggestions as to the kinds of 'approximate' rationality that might be employed by an organism possessing limited information and limited computational facilities. The suggestions were 'hypothetical' in that, lacking definitive knowledge of the human decisional processes, we can only conjecture on the basis of our everyday experiences, our introspection, and a very limited body of psychological literature what these processes are. The suggestions were intended, however, as empirical statements, however tentative, about some of the actual mechanisms involved in human and other organismic choice.[1]

Now if an organism is confronted with the problem of behaving approximately rationally, or adaptively, in a particular environment, the kinds of simplifications that are suitable may depend not only on the characteristics – sensory, neural, and other – of the organism, but equally upon the structure of the environment. Hence, we might hope to discover, by a careful examination of some of the fundamental structural characteristics of the environment, some further clues as to the nature of the approximating mechanisms used in decision making. This is the line of attack that will be adopted in the present paper.

The environment we shall discuss initially is perhaps a more appropriate one for a rat than for a human. For the term *environment* is ambiguous. We are not interested in describing some physically objective world in its totality, but only those aspects of the totality that have relevance as the 'life space' of the organism considered. Hence, what we call the 'environment' will depend upon the 'needs', 'drives', or 'goals' of the organism, and upon its perceptual apparatus.

1. Since writing the paper referred to I have found confirmation for a number of its hypotheses in the interesting and significant study, by A. de Groot, of the thought processes of chess players (3). I intend to discuss the implications of these empirical findings for my model in another place.

The Environment of the Organism

We consider first a simplified (perhaps 'simple-minded') organism that has a single need – food – and is capable of three kinds of activity: resting, exploration, and food getting. The precise nature of these activities will be explained later. The organism's life space may be described as a surface over which it can locomote. Most of the surface is perfectly bare, but at isolated, widely scattered points there are little heaps of food, each adequate for a meal.

The organism's vision permits it to see, at any moment, a circular portion of the surface about the point in which it is standing. It is able to move at some fixed maximum rate over the surface. It metabolizes at a given average rate and is able to store a certain amount of food energy, so that it needs to eat a meal at certain average intervals. It has the capacity, once it sees a food heap, to proceed toward it at the maximum rate of locomotion. The problem of rational choice is to choose its path in such a way that it will not starve.

Now I submit that a rational way for the organism to behave is the following: (a) it explores the surface at random, watching for a food heap; (b) when it sees one, it proceeds to it and eats (food getting); (c) if the total consumption of energy during the average time required, per meal, for exploration and food getting is less than the energy of the food consumed in the meal, it can spend the remainder of its time in resting.[2]

There is nothing particularly remarkable about this description of rational choice, except that it differs so sharply from the more sophisticated models of human rationality that have been proposed by economists and others. Let us see what it is about the organism and its environment that makes its choice so simple.

1. It has only a single goal: food. It does not need to weigh the

2. A reader who is familiar with W. Grey Walter's mechanical turtle, *Machina speculatrix* (8), will see as we proceed that the description of our organism could well be used as a set of design specifications to assure the survival of his turtle in an environment sparsely provided with battery chargers. Since I was not familiar with the structure of the turtle when I developed this model, there are some differences in their behavior – but the resemblance is striking.

respective advantages of different goals. It requires no 'utility function' or set of 'indifference curves' to permit it to choose between alternatives.

2. It has no problem of maximization. It needs only to maintain a certain average rate of food intake, and additional food is of no use to it. In the psychologist's language, it has a definite, fixed aspiration level, and its successes or failures do not change its aspirations.

3. The nature of its perceptions and its environment limit sharply its planning horizon. Since the food heaps are distributed randomly, there is no need for pattern in its searching activities. Once it sees a food heap, it can follow a definite 'best' path until it reaches it.

4. The nature of its needs and environment create a very natural separation between 'means' and 'ends'. Except for the food heaps, one point on the surface is as agreeable to it as another. Locomotion has significance only as it is a means to reaching food.[3]

We shall see that the first point is not essential. As long as aspirations are fixed, the planning horizon is limited, and there is a sharp distinction between means and ends, the existence of multiple goals does not create any real difficulties in choice. The real complications ensue only when we relax the last three conditions; but to see clearly what is involved, we must formulate the model a little more precisely.

Perceptual Powers, Storage Capacity, and Survival

It is convenient to describe the organism's life space not as a continuous surface, but as a branching system of paths, like a maze, each branch point representing a choice point. We call the selection of a branch and locomotion to the next branch point a 'move'. At a small fraction of the branch points are heaps of food.

Let p, $0 < p < 1$, be the percentage of branch points, randomly

3. It is characteristic of economic models of rationality that the distinction between 'means' and 'ends' plays no essential role in them. This distinction *cannot* be identified with the distinction between behavior alternatives and utilities, for reasons that are set forth at some length in (5, ch. 4 and 5).

distributed, at which food is found. Let d be the average number of paths diverging from each branch point. Let v be the number of moves ahead the organism can see. That is, if there is food at any of the branch points within v moves of the organism's present position, it can select the proper paths and reach it. Finally let H be the maximum number of moves the organism can make between meals without starving.

At any given moment, the organism can see d branch points at a distance of one move from his present position, d^2 points two moves away, and in general, d^k points k moves away. In all, it can see

$$d + d^2 + \ldots + d^v = \frac{d}{d-1}(d^v - 1)$$

points. When it chooses a branch and makes a move, d^v new points become visible on its horizon. Hence, in the course of m moves, md^v new points appear. Since it can make a maximum of H moves, and since v of these will be required to reach food that it has discovered on its horizon, the probability, $Q = 1 - P$, that it will *not* survive will be equal to the probability that no food points will be visible in $(H - v)$ moves. (If p is small, we can disregard the possibility that food will be visible inside its planning horizon on the first move.) Let ρ be the probability that none of the d^v new points visible at the end of a particular move is a food point.

$$\rho = (1 - p)d^v. \qquad \textbf{1.1}$$

Then:

$$1 - P = Q = \rho^{(H-v)} = (1 - p)^{(H-v)d^v}. \qquad \textbf{1.2}$$

We see that the survival chances, from meal to meal, of this simple organism depend on four parameters, two that describe the organism and two the environment: p, the richness of the environment in food; d, the richness of the environment in paths; H, the storage capacity of the organism; and v, the range of vision of the organism.

To give some impression of the magnitudes involved, let us assume that p is 1/10,000, $(H - v)$ is 100, d is 10 and v is 3. Then the probability of seeing a new food point after a move is $1 - \rho = 1 - (1 - p)^{1000} \sim 880/10{,}000$, and the probability of survival is $P = 1 - \rho^{100} \sim 9999/10{,}000$. Hence there is in this case only one chance in 10,000 that the organism will fail to reach a food point before the end of the survival interval. Suppose now that the survival time $(H - v)$ is increased one third, that is, from 100 to 133. Then a similar computation shows that the chance of starvation is reduced to less than one chance in 100,000. A one-third increase in v will, of course, have an even greater effect, reducing the chance of starvation from one in 10^{-4} to one in 10^{-40}.

Using the same values, $p = 0{\cdot}0001$, and $(H - v) = 100$, we can compute the probability of survival if the organism behaves completely randomly. In this case $P' = [1 - (1 - p)^{100}] = 000{\cdot}9$. From these computations, we see that the organism's modest capacity to perform purposive acts over a short planning horizon permits it to survive easily in an environment where random behavior would lead to rapid extinction. A simple computation shows that its perceptual powers multiply by a factor of 880 the average speed with which it discovers food.

If p, d, and v are given, and in addition we specify that the survival probability must be greater than some number close to unity $(P \geqslant 1 - \epsilon)$, we can compute from **1.2** the corresponding minimum value of H:

$$\log (1 - P) = (H - v) \log \rho \qquad \textbf{1.3}$$

$$H \geqslant v + \frac{\log \epsilon}{\log \rho} \qquad \textbf{1.4}$$

For example, if $\rho = 0{\cdot}95$ and $\epsilon = 10^{-10}$, then $\log \rho = -0{\cdot}022$, $\log \epsilon = -10$ and $(H - v) \geqslant 455$. The parameter, H, can be interpreted as the 'storage capacity' of the organism. That is, if the organism metabolizes at the rate of α units per move, then a storage of αH food units, where H is given by **1.4**, would be required to provide survival at the specified risk level, ϵ.

Further insight into the meaning of H can be gained by considering the average number of moves, M, required to discover food. From **1.1**, the probability of making $(k - 1)$ moves without discovering food, and then discovering it on the kth is:

$$P_k = (1 - \rho)\rho^{(k-1)} \qquad \textbf{1.5}$$

Hence, the average number of moves, M, required to discover food is:

$$M = \sum_{k=1}^{\infty} k(1-\rho)\rho^{k-1}$$

$$= \frac{(1-\rho)}{(1-\rho)^2} = \frac{1}{(1-\rho)} \qquad \textbf{1.6}$$

Since $(1-\rho)$ is the probability of discovering food in any one move, M is the reciprocal of this probability. Combining **1.3** and **1.6**, we obtain:

$$\frac{M}{H-v} = \frac{\log \rho}{(1-\rho)} \frac{1}{\log (1-P)} \qquad \textbf{1.7}$$

Since ρ is close to one, $\log_e \rho \simeq (1-\rho)$, and **1.7** reduces approximately to:

$$\frac{M}{H-v} \simeq \frac{1}{\log_e (1-P)} \qquad \textbf{1.8}$$

For example, if we require $(1-P) = \epsilon \leqslant 10^{-4}$ (one chance in 10,000 of starvation), then $M/(H-v) \leqslant 0{\cdot}11$. For this survival level we require food storage approximately equal to $\alpha(v + 9M)$ – food enough to sustain the organism for nine times the period required, on the average, to discover food, plus the period required to reach the food.[4]

Choice Mechanisms for Multiple Goals

We consider now a more complex organism capable of searching for and responding to two or more kinds of goal objects. In doing this we could introduce any desired degree of complexity into the choice process; but the interesting problem is how to introduce multiple goals with a minimum complication of the process – that

4. I have not discovered any very satisfactory data on the food storage capacities of animals, but the order of magnitude suggested above for the ratio of average search time to storage capacity is certainly correct. It may be noted that, in some cases at least, where the 'food' substance is ubiquitous, and hence the search time negligible, the storage capacity is also small. Thus, in terrestrial animals there is little oxygen storage and life can be maintained in the absence of air for only a few minutes. I am not arguing as to which way the causal arrow runs, but only that the organisms, in this respect, are adapted to their environments and do not provide storage that is superfluous.

is, to construct an organism capable of handling its decision problems with relatively primitive choice mechanisms.

At the very least, the presence of two goals will introduce a consistency requirement – the time consumed in attaining one goal will limit the time available for pursuit of the other. But in an environment like the one we have been considering, there need be no further relationship between the two goals. In our original formulation, the only essential stipulation was that H, the storage capacity, be adequate to maintain the risk of starvation below a stipulated level $(1 - P)$. Now we introduce the additional stipulation that the organism should only devote a fraction, λ, of its time to food-seeking activities, leaving the remaining fraction, $1 - \lambda$, to other activities. This new stipulation leads to a requirement of additional storage capacity.

In order to control the risk of starving, the organism must begin its exploration for food whenever it has reached a level of H periods of food storage. If it has a total storage of $(\mu + H)$ periods of food, and if the food heaps are at least $\alpha(\mu + H)$ in size, then it need begin the search for food only μ periods after its last feeding. But the food research will require, on the average, M periods. Hence, if a hunger threshold is established that leads the organism to begin to explore μ periods after feeding, we will have:

$$\lambda = \frac{M}{M + \mu}. \qquad 2.1$$

Hence, by making μ sufficiently large, we can make λ as small as we please. Parenthetically, it may be noted that we have here a close analogue to the very common two-bin system of controlling industrial inventories. The primary storage, H, is a buffer stock to meet demands pending the receipt of new orders (with risk, $1 - P$, of running out); the secondary storage, μ, defines the 'order point'; and $\mu + M$ is the average order quantity. The storage μ is fixed to balance storage 'costs' against the cost (in this case, time pressure) of too frequent reordering.

If food and the second goal object (water, let us say) are randomly and independently distributed, then there are no important complications resulting from interference between the two activities. Designate by the subscript 1 the variables and parameters referring to food getting (e.g. μ_1 is the food threshold in periods), and by the subscript 2 the quantities referring to water seeking. The organism will have adequate time for both activities if $\lambda_1 + \lambda_2 < 1$.

Now when the organism reaches either its hunger or thirst threshold, it will begin exploration. We assume that if *either* of the goal objects becomes visible, it will proceed to that object and satisfy its hunger or thirst (this will not increase the number of moves required, on the average, to reach the other object); but if *both* objects become visible at the same time, and if S_1 and S_2 are the respective quantities remaining in storage at this time, then it will proceed to food or water as M_1/S_1 is greater or less than M_2/S_2. This choice will maximize its survival probability. What is required, then, is a mechanism that produces a drive proportional to M_i/S_i.

A priority mechanism of the kind just described is by no means the only or simplest one that can be postulated. An even simpler rule is for the organism to persist in searching for points that will satisfy the particular need that first reached its threshold and initiated exploratory behavior. This is not usually an efficient procedure, from the standpoint of conserving goal-teaching time, but it may be entirely adequate for an organism generously endowed with storage capacity.

We see that an organism can satisfy a number of distinct needs without requiring a very elaborate mechanism for choosing among them. In particular, we do not have to postulate a utility function or a 'marginal rate of substitution'.

We can go even further, and assert that a primitive choice mechanism is adequate to take advantage of important economies, if they exist, which are derivable from the interdependence of the activities involved in satisfying the different needs. For suppose the organism has n needs, and that points at which he can satisfy each are distributed randomly and independently through the environment, each with the same probability, p. Then the probability that no points satisfying *any* of the needs will be visible on a particular move is ρ^n, and the mean number of moves for discovery of the *first* need-satisfying point is:

$$m_n = \frac{1}{(1 - \rho^n)} \qquad 2.2$$

Suppose that the organism begins to explore, moves to the first need-satisfying point it discovers, resumes its exploration, moves to the next point it discovers that satisfies a need other than the one already satisfied, and so on. Then the mean time required to search for all n goals will be:

$$M_n = m_n + m_{n-1} + \ldots = \sum_{i=1}^{n} \frac{1}{(1-\rho^i)} \leqslant \frac{n}{(1-\rho)} \qquad \textbf{2.3}$$

In particular, if ρ is close to one, that is, if need-satisfying points are rare, we will have:

$$M_n - M_{n-1} = \frac{1}{(1-\rho^n)} = \frac{1}{(1-\rho)} \times \frac{1}{\sum_{i=0}^{n} \rho^i} \simeq \frac{M_1}{n} \qquad \textbf{2.4}$$

and

$$M_n \simeq M_1 \sum_{i=1}^{n} \frac{1}{i} \qquad \textbf{2.5}$$

Now substituting particular values for n in **2.5** we get: $M_2 = 3/2\, M_1$, $M_3 = 11/6\, M_1$; $M_4 = 25/12\, M_1$, etc. We see that if the organism has two separate needs, its exploration time will be only 50 per cent greater than – and not twice as great as – if it has only one need; for four needs the exploration time will be only slightly more than twice as great as for a single need, and so on. A little consideration of the program just described will show that the joint exploratory process does not reduce the primary storage capacity required by the organism but does reduce the secondary storage capacity required. As a matter of fact, there would be no necessity at all for secondary storage.

This conclusion holds only if the need-satisfying points are *independently* distributed. If there is a negative correlation in the joint distribution of points satisfying different needs, then it may be economical for the organism to pursue its needs separately, and hence to have a simple signalling mechanism, involving secondary storage, to trigger its several exploration drives. This point will be developed further in the next section.

A word may be said here about 'avoidance needs'. Suppose that certain points in the organism's behavior space are designated as 'dangerous'. Then it will need to avoid those paths that lead to these particular points. If a fraction r of all points, randomly distributed, are dangerous, then the number of available paths, among those visible at a given move, will be reduced to $(1 - r)d^v$. Hence, $\rho' = (1 - p)^{(1-r)dv}$ will be smaller than ρ (equation **1.1**), and M (equation **1.6**) will be correspondingly larger. Hence, the presence of danger points simply increases the average exploration time and, consequently, the required storage capacity of the organism.

Further Specification of the Environment: Clues

In our discussion up to the present point, the range of the organism's anticipations of the future has been limited by the number of behavior alternatives available to it at each move (d), and the length of the 'vision' (v). It is a simple matter to introduce into the model the consequences of several types of learning. An increase in the repertoire of behavior alternatives or in the length of vision can simply be represented by changes in d and v, respectively.

A more interesting possibility arises if the food points are not distributed completely at random, and if there are clues that indicate whether a particular intermediate point is rich or poor in paths leading to food points. First, let us suppose that on the path leading up to each food point the k preceding choice points are marked with a food clue. Once the association between the clue and the subsequent appearance of the food point is learned by the organism, its exploration can terminate with the discovery of the clue, and it can follow a determinate path from that point on. This amounts to substituting $v' = (v + k)$ for v.

A different kind of clue might operate in the following fashion. Each choice point has a distinguishable characteristic that is associated with the probability of encountering a food point if a path is selected at random leading out of this choice point. The organism can then select at each move the choice point with the highest probability. If only certain choice points are provided with such clues, then a combination of random and systematic exploration can be employed. Thus the organism may be led into 'regions' where the probability of goal attainment is high relative to other regions, but it may have to explore randomly for food within a given region.

A concrete example of such behavior in humans is the 'position play' characteristic of the first phase of a chess game. The player chooses moves on the basis of certain characteristics of resulting positions (e.g. the extent to which his pieces are developed). Certain positions are adjudged richer in attacking and defensive possibilities than others, but the original choice may involve no definite plan for the subsequent action after the 'good' position has been reached.

Next, we turn to the problem of choice that arises when those regions of the behavior space that are rich in points satisfying one need (p_1 is high in these regions) are poor in points satisfying another need (p_2 is low in these same regions). In the earlier case of goal conflict (two or more points simultaneously visible mediating different needs), we postulated a priority mechanism that amounted to a mechanism for computing relative need intensity and for responding to the more intense need. In the environment with clues, the learning process would need to include a conditioning mechanism that would attach the priority mechanism to the response to competing clues, as well as to the response to competing visible needs.

Finally, we have thus far specified the environment in such a way that there is only one path leading to each point. Formally, this condition can always be satisfied by representing as two or more points any point that can be reached by multiple paths. For some purposes, it might be preferable to specify an environment in which paths converge as well as diverge. This can be done without disturbing the really essential conditions of the foregoing analysis. For behavior of the sort we have been describing, we require of the environment only:

1. that if a path is selected *completely* at random the probability of survival is negligible;

2. that there exist clues in the environment (either the actual visibility of need-satisfying points or anticipatory clues) which permit the organism, sufficiently frequently for survival, to select specific paths that lead with certainty, or with very high probability, to a need-satisfying point.

Concluding Comments on Multiple Goals

The central problem of this paper has been to construct a simple mechanism of choice that would suffice for the behavior of an organism confronted with multiple goals. Since the organism, like those of the real world, has neither the senses nor the wits to discover an 'optimal' path – even assuming the concept of optimal to be clearly defined – we are concerned only with finding a choice mechanism that will lead it to pursue a 'satisficing' path, a path that will permit satisfaction at some specified level of all of its needs.

Certain of the assumptions we have introduced to make this possible represent characteristics of the organism.

(a). It is able to plan short purposive behavior sequences (of length not exceeding v), but not long sequences.
(b). Its needs are not insatiable, and hence it does not need to balance marginal increments of satisfaction. If all its needs are satisfied, it simply becomes inactive.
(c). It possesses sufficient storage capacity so that the exact moment of satisfaction of any particular need is not critical.

We have introduced other assumptions that represent characteristics of the environment, the most important being that need satisfaction can take place only at 'rare' points which (with some qualifications we have indicated) are distributed randomly.

The most important conclusion we have reached is that blocks of the organism's time can be allocated to activities related to individual needs (separate means–end chains) without creating any problem of over-all allocation or coordination or the need for any general 'utility function'. The only scarce resource in the situation is time, and its scarcity, measured by the proportion of the total time that the organism will need to be engaged in *some* activity, can be reduced by the provision of generous storage capacity.

This does not mean that a more efficient procedure cannot be constructed, from the standpoint of the total time required to meet the organism's needs. We have already explored some simple possibilities for increasing efficiency by recognizing complementarities among activities (particularly the exploration activity). But the point is that these complications are not essential to the survival of an organism. Moreover, if the environment is so constructed (as it often is in fact) that regions rich in possibilities for one kind of need satisfaction are poor in possibilities for other satisfactions, such efficiencies may not be available.

It may be objected that even relatively simple organisms appear to conform to efficiency criteria in their behavior, and hence that their choice mechanisms are much more elaborate than those we have described. A rat, for example, learns to take shorter rather than longer paths to food. But this observation does not affect the central argument. We can introduce a mechanism that leads the organism to choose time-conserving paths, where multiple paths are available for satisfying a given need, without any assumption of a mechanism that allocates time among *different* needs. The former mechanism simply increases the 'slack' in the whole system, and makes it even more feasible to ignore the complementarities among activities in programming the over-all behavior of the organism.

This is not the place to discuss at length the application of the model to human behavior, but a few general statements may be in order. First, the analysis has been a static one, in the sense that we have taken the organism's needs and its sensing and planning capacities as given. Except for a few comments, we have not considered how the organism develops needs or learns to meet them. One would conjecture, from general observation and from experimentation with aspiration levels, that in humans the balance between the time required to meet needs and the total time available is maintained by the raising and lowering of aspiration levels. I have commented on this point at greater length in my previous paper.[5]

Second, there is nothing about the model that implies that the needs are physiological and innate rather than sociological and acquired. Provided that the needs of the organism can be specified at any given time in terms of the aspiration levels for the various kinds of consummatory behavior, the model can be applied.

5. See (6, pp. 111, 117–18). For an experiment demonstrating the adjustment of the rat's aspiration levels to considerations of realizability, see Festinger (2).

The principal positive implication of the model is that we should be skeptical in postulating for humans, or other organisms, elaborate mechanisms for choosing among diverse needs. Common denominators among needs may simply not exist, or may exist only in very rudimentary form; and the nature of the organism's needs in relation to the environment may make their nonexistence entirely tolerable.

There is some positive evidence bearing on this point in the work that has been done on conflict and frustration. A common method of producing conflict in the laboratory is to place the organism in a situation where: (a) it is stimulated to address itself simultaneously to alternative goal-oriented behaviors, or (b) it is stimulated to a goal-oriented behavior, but restricted from carrying out the behaviors it usually evinces in similar natural situations. This suggests that conflict may arise (at least in a large class of situations) from presenting the animal with situations with which it is not 'programmed' to deal. Conflict of choice may often be equivalent to an absence of a choice mechanism in the given situation. And while it may be easy to create such situations in the laboratory, the absence of a mechanism to deal with them may simply reflect the fact that the organism seldom encounters equivalent situations in its natural environment (see 4, ch. 14).

Conclusion

In this paper I have attempted to identify some of the structural characteristics that are typical of the 'psychological' environments of organisms. We have seen that an organism in an environment with these characteristics requires only very simple perceptual and choice mechanisms to satisfy its several needs and to assure a high probability of its survival over extended periods of time. In particular, no 'utility function' needs to be postulated for the organism, nor does it require any elaborate procedure for calculating marginal rates of substitution among different wants.

The analysis set forth here casts serious doubt on the usefulness of current economic and statistical theories of rational behavior as bases for explaining the characteristics of human and other organismic rationality. It suggests an alternative approach to the description of rational behavior that is more closely related to psychological theories of perception and cognition, and that is in closer agreement with the facts of behavior as observed in laboratory and field.

References

1. W. Edwards, 'The theory of decision making', *Psychological Bulletin*, vol. 51 (1954), pp. 380–417.
2. L. Festinger, 'Development of differential appetite in the rat', *Journal of Experimental Psychology*, vol. 32 (1953), pp. 226–34.
3. A. de Groot, *Het Denken van den Schaker*, Noord-Hollandsche Uitgevers Maatschapij, 1946.
4. J. McV. Hunt, *Personality and the Behavior Disorders*, Ronald, 1944.
5. H. A. Simon, *Administrative Behavior*, Macmillan, 1941.
6. H. A. Simon, 'A behavioral model of rational choice', *Quarterly Journal Economics*, vol. 59 (1955), pp. 99–118.
7. R. M. Thrall, C. H. Coombs, and R. L. Davis (eds.), *Decision Processes*, Wiley, 1954.
8. W. G. Walter, *The Living Brain*, Norton, 1953.

11 W. R. Ashby

Adaptation in the Multistable System

Excerpt from chapter 16 of W. R. Ashby, *Design for a Brain*, Wiley, 2nd edn, 1960, pp. 205–14.

1.1. Continuing our study of types of environment we next consider, after Figures 15/7/1 and 15/8/1 [not included in this excerpt], the case in which the subsystems of the environment are connected unrestrictedly in direction, so that feedbacks occur between them. This type of environment may vary according to the *amounts* of communication (variety) that are transmitted between subsystem and subsystem. Two degrees are of special interest as types:

1. Those in which it is near the maximum – the richly joined environment. (The exposition is more convenient if we consider this case first, as it can be dismissed briefly.)
2. Those in which the amount is small.

The Richly Joined Environment

1.2. When a set of subsystems is richly joined, each variable is as much affected by variables in other subsystems as by those in its own. When this occurs, the division of the whole into subsystems ceases to have any natural basis.

The case of the richly joined environment thus leads us back to the case discussed in Chapter 11 [not included in this excerpt].

1.3. Examples of environments that are both large and richly connected are not common, for our terrestrial environment is widely characterized by being highly subdivided. A richly connected environment would therefore intuitively be perceived as something unusual, or even unnatural. The examples given below are somewhat *recherché*, but they will suffice to make clear what is to be expected in this case.

The combination lock was mentioned in section 11/9 [not

included here]. Though not vigorous dynamically, its parts, so far as they affect the output at the bolt, are connected in that the relations between them are highly conditional. Thus, if there are seven dials that allow the bolt to move only when set at RHOMBUS, then the effect of the first dial going to R on the movement of the bolt is conditional on the positions of all the other six; and similarly for the second and remaining letters.

A second example is given by a set of simultaneous equations, which can legitimately be regarded as the temporary environment of a professional computer if he is paid simply to get correct answers. Sometimes they come in the simplest form, e.g.

$$\left.\begin{aligned} 2x &= 8 \\ 3y &= -7 \\ \tfrac{1}{2}z &= 3 \end{aligned}\right\}$$

Then they correspond to the iterated form; and each line can be treated without reference to the others, as in section 15/4 [not included here].

Sometimes they are rather more complex, e.g.

$$\left.\begin{aligned} 2x \qquad\quad &= 3 \\ 3x - 2y \quad &= 2 \\ x + y - z &= 0 \end{aligned}\right\}$$

This form can be solved serially, as in section 15/8 [not included here][1]; for the first line can be treated without reference to the other two; then when the first process has been successful the second line can be treated without reference to the last; and so to the end. The peculiarity of this form is that the value of x is *not conditional* on the values of the coefficients in the second and third lines.

Sometimes the forms are more complex, e.g.

$$\left.\begin{aligned} 2x + y - 3z &= 2 \\ x - y + 2z &= 0 \\ -x - 3y + z &= 1 \end{aligned}\right\}$$

1. Reference is often made to figures and sections from other chapters of *Design for a Brain*. The volume itself should be consulted if further information is required – *editor*.

Now the value of x is conditional on the values of *all* the coefficients; and in finding x, no coefficient can be ignored. The same is true of y and z. Thus if we regard the coefficients as the environment and the values of x, y and z as output, correctness in the answer demands that, in getting any part of the answer (any one of the three values), *all* the environment must be taken into account.

A third, and more practical, example of a richly connected environment (now, thank goodness, no more) faced the experimenter in the early days of the cathode-ray oscilloscope. Adjusting the first experimental models was a matter of considerable complexity. An attempt to improve the brightness of the spot might make the spot also move off the screen. The attempt to bring it back might alter its rate of sweep and start it oscillating vertically. An attempt to correct this might make its line of sweep leave the horizontal; and so on. This system's variables (brightness of spot, rate of sweep, etc.) were dynamically linked in a rich and complex manner. Attempts to control it through the available parameters were difficult precisely because the variables were richly joined.

1.4. How long will an ultrastable system that includes such an environment take to get adapted?

This is the question of section 11/2. Unless a large fraction of the outcomes are acceptable, the time taken tends to be like T_1 of section 11/5. As the system is made larger, so does the time of adaptation tend to increase beyond all bounds of what is practical; in other words, the ultrastable system probably fails. But this failure does not discredit the ultrastable system as a model of the brain (section 8/17), for such an environment is one that is also likely to defeat the living brain. That the living organism is notoriously apt to find such environments difficult or impossible for adaptation is precisely the reason why the combination lock is relied on for protection.

Even when a skilled thief defeats the combination lock he supports, rather than refutes, the thesis. Thus if he can hear, as each dial moves, a tumbler fall into position, then the environment is to him a serial one (section 15/8); for he can get the first dial-setting right without reference to the others, then the second, and so on. The time of its opening is thus made vastly shorter. Thus the skilled thief does not really adapt successfully to the richly joined environment – he demonstrates that what to others is richly joined is to him joined serially.

Thus the first answer to the question – how does the ultrastable system, or the brain, adapt to a richly joined environment – is, it doesn't. After the reasonableness of this answer has been made clear, we may then notice that sometimes there are ways in which an environment, apparently too complex for adaptation, may eventually be adapted to; perhaps by the discovery of ways of getting through the necessary trials much faster, or perhaps by the discovery that the environment is not really as complex as it looks. (*I* to *C*, section 13/4, discusses the matter.)

The Poorly Joined Environment

1.5. We will finally consider the case in which the environment consists of subsystems joined so that they affect one another only weakly, or occasionally, or only through other subsystems. It was suggested in section 15/2 that this is the common case in almost all natural terrestrial environments.

If the degree of interaction between the subsystems varies, its limits are: at the lower end, the iterated systems of section 15/4 (as the communication between subsystems falls to zero), and at the upper end, the richly connected systems of section 16/2 (as the communication rises to its maximum).

When the communication *between* subsystems falls much below that *within* subsystems, the subsystems will show naturally and prominently (section 12/17).

If such an environment acts within an ultrastable system, what will happen? Will adaptation occur? As the discussion below will show, the number of cases is so many, and the forms so various, that no detailed and exhaustive account is possible. We must therefore use the strategy of section 2/17, getting certain type-forms quite clear, and then covering the remainder by some appeal to continuity: that so far as other forms resemble the type-forms in their construction, to a somewhat similar degree will they resemble the type-forms in their behavior.

1.6. To obtain a secure basis for the discussion of this most important case, let us state explicitly what is now assumed:

The Environment of a System

1. The environment is assumed to be as described in section 15/2, so that it consists of large numbers of subsystems that have many states of equilibrium. The environment is thus assumed to be polystable.
2. Whether because the primary joins between the subsystems are few, or because equilibria in the subsystems are common, the interaction between subsystems is assumed to be weak.
3. The organism coupled to this environment will adapt by the basic method of ultrastability, i.e. by providing second-order feedbacks that veto all states of equilibrium except those that leave each essential variable within its proper limits.
4. The organism's reacting part is itself divided into subsystems between which there is no direct connexion. Each subsystem is assumed to have its own essential variables and second-order feedback. Figure 1 illustrates the connexions, but somewhat inadequately, for it shows only three subsystems. (It should be compared with Figures 15/7/1 and 15/8/1.)

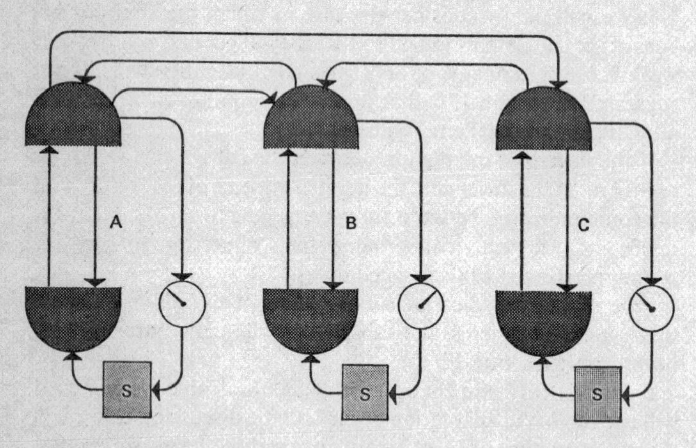

Figure 1

Such a system is essentially similar to the *multistable* system defined in the first edition. (The system defined there allowed more freedom in the connexions between the main variables, e.g. from reacting part to reacting part, and between reacting part and an environmental subsystem other than that chiefly joined to it; these minor variations are a nuisance and of little importance – in the next chapter we shall be considering such variations.)

1.7. To trace the behavior of the multistable system, suppose that we are observing two of the subsystems, e.g. A and B of Figure 1, that their main variables are directly linked so that changes of either immediately affects the other, and that for some reason all the other subsystems are inactive.

The first point to notice is that, as the other subsystems are inactive, their presence may be ignored; for they become like the 'background' of section 6/1. Even if some are active, they can still be ignored if the two observed subsystems are separated from them by a wall of inactive subsystems (section 12/10).

The next point to notice is that the two subsystems, regarded as a unit, *form a whole which is ultrastable*. This whole will therefore proceed, through the usual series of events, to a terminal field. Its behavior will not be essentially different from that recorded in Figure 8/4/1. If, however, we regard the same series of events as occurring, not within one ultrastable whole, but as interactions between a minor environment and a minor organism, each of two subsystems, then we shall observe behaviors homologous with those observed when interaction occurs between 'organism' and 'environment'. In other words, *within a multistable system, subsystem adapts to subsystem in exactly the same way as animal adapts to environment.* Trial and error will appear to be used; and, when the process is completed, the activities of the two parts will show coordination to the common end of maintaining the essential variables of the double system within their proper limits.

Exactly the same principle governs the interactions between three subsystems. If the three are in continuous interaction, they form a single ultrastable system which will have the usual properties.

As illustration we can take the interesting case in which two of them, A and C say, while having no immediate connexion with each other, are joined to an intervening system B, intermittently but not simultaneously. Suppose B interacts first with A: by their ultrastability they will arrive at a terminal field. Next let B and C interact. If B's step-mechanisms, together with those of C, give a stable field to the main variables of B and C, then that set of B's step-mechanism values will persist indefinitely; for when B rejoins A the original stable field will be re-formed. But if B's set with C's does not give stability, then it will be changed to another set. It follows that B's step-mechanisms will stop changing when, and only when, they have a set of values which forms fields stable with both A and C. (The identity in principle with the process described in section 8/10 should be noted.)

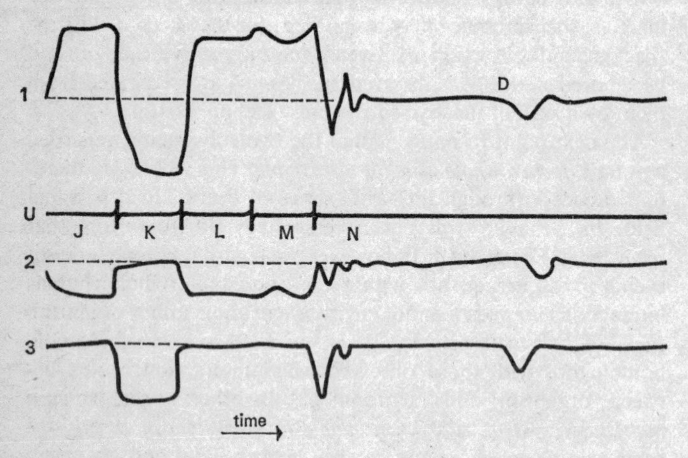

Figure 2 Three units of the Homeostat interacting. Bars in the central positions prevent 2 and 3 from moving in the direction corresponding here to upwards. Vertical strokes on U record changes of uniselector position in unit 1. Disturbance D, made by the operator, demonstrates the whole's stability

1.8. The process can be illustrated on the Homeostat. Three units were connected so that the diagram of immediate effects was $2 \rightleftarrows 1 \rightleftarrows 3$ (corresponding to A, B, and C respectively). To separate the effects of 2 and 3 on 1, bars were placed across the potentiometer dishes (Figure 8/2/2) of 2 and 3 so that they could move only in the direction recorded as downwards in Figure 1, while 1 could move either upwards or downwards. If 1 was above the central line (shown broken), 1 and 2 interacted, and 3 was independent; but if 1 was below the central line, then 1 and 3 interacted, and 2 was independent. 1 was set to act on 2 negatively and on 3 positively, while the effects $2 \longrightarrow 1$ and $3 \longrightarrow 1$ were uniselector-controlled.

When switched on, at J, 1 and 2 formed an unstable system and the critical state was transgressed. The next uniselector connexions (K) made 1 and 2 stable, but 1 and 3 were unstable. This led to the next position (L) where 1 and 3 were stable but 1 and 2 became again unstable. The next position (M) did not remedy this; but the following position (N) happened to provide connexions which made both systems stable. The values of the step-mechanisms are now permanent; 1 can interact repeatedly with both 2 and 3 without loss of stability.

It has already been noticed that if A, B and C should form from time to time a triple combination, then the step-mechanisms of all three parts will stop changing when, and only when, the triple combination has a stable field. But we can go further than that. If A, B and C should join intermittently in various ways, sometimes joining as pairs, sometimes as a triple, and sometimes remaining independent, then their step-mechanisms will stop changing when, and only when, they arrive at a set of values which gives stability to *all* the arrangements.

Clearly the same line of reasoning will apply no matter how many subsystems interact or in what groups or patterns they join. Always we can predict that *their step-mechanisms will stop changing when, and only when, the combinations are all stable*. Ultrastable systems, whether isolated or joined in multistable systems, act always selectively towards those step-mechanism values which provide stability.

1.9. At the beginning of the preceding section it was assumed, for simplicity, that the process of dispersion was suspended, for we assumed that the two subsystems interacting remained the same two (e.g. A and B of Figure 1) during the whole process. What modifications must be made when we allow for the fact that in a multistable system the number and distribution of subsystems active at each moment may fluctuate by dispersion?

The progression to equilibrium of the whole, to a terminal field, and thus to adaptation of the whole, will occur whether dispersion occurs or not. The effect of dispersion is to destroy the individuality of the subsystems considered in the previous section. There two subsystems were pictured as going through the complex processes of ultrastability, their main variables being repeatedly active while those of the surrounding subsystems remained inactive. This permanence of individuality can hardly occur when dispersion occurs. Thus, suppose that a multistable system's field of all its main variables is stable, and that its representative point is at a state of equilibrium R. If the representative point is displaced to a point P, the lines from this point will lead it back to R. As the point travels back from P to R, subsystems will come into action, perhaps singly, perhaps in combination, becoming active and inactive in kaleidoscopic variety and apparent confusion. Travel along another line to R will also activate various combinations of subsystems; and the set made active in the second line may be very different from that made active by the first.

In such conditions it is no longer profitable to observe particular subsystems when a multistable system adapts. What will happen is that so long as some essential variables are outside their limits, so long will change at step-mechanisms cause combination after combination of subsystems to become active. But when a stable field arises not causing step-mechanisms to change, it will, as usual, be retained. If now the multistable system's adaptation be tested by displacements of its representative point, the system will be found to respond by various activities of various subsystems, all coordinated to the common end. But though coordinated in this way, there will, in general, be no simple relation between the actions of subsystem on subsystem: knowing which subsystems were activated on one line of behavior, and how they interacted, gives no certainty about which will be activated on some other line of behavior, or how they will interact.

Later I shall refer again to 'subsystem A adapting to, or interacting with, subsystem B', but this will be only a form of words, convenient for description: it is to be understood that what is A and what is B may change from moment to moment.

1.10. This new picture answers the objection to Figure 1 (and the others) that it shows a tidiness nowhere evident when the organism (or the environment) is examined anatomically or histologically. The figures are diagrams of immediate effects, and are intended purely as an aid to easier thinking about functional and behavioral relationships. They must be regarded as showing only *functional* connexions, and of these only those between variables that are active over some small interval of time. Figure 1 is thus apt to mislead both by suggesting a permanence of structure that does not exist when dispersion occurs, and by suggesting an actual two-dimensional form that may well have no anatomical or histological existence. Nevertheless, the functional relationships are indisputable, and the figure represents them. How they are related to variables physically identifiable in the brain has yet to be discovered.

1.11. Though the multistable system may look chaotic in action, as the activity fluctuates over the subsystems with the same apparent lack of order as that shown by the smoking chimneys of section 13/18, yet the tendency is always towards ultimate equilibrium and adaptation. So the next question to ask is that of Chapter 11: will the adaptation take an excessively long time?

Clearly, following the arguments of the previous chapter, much will depend on the richness of connexion between the subsystems – on how much disturbance comes to each subsystem from the others.

At the limit, when the transfers of disturbance are all zero, the whole system becomes identical with the iterated systems of section 15/6, and the whole will progress to adaptation similarly. In this case the time taken to reach adaptation will be the moderate time of T_3, rather than the excessive time of T_1.

As the connexions become richer, whether by more basic joins or by the subsystems having fewer states of equilibrium, so will the system move towards the richly connected type of section 16/4; and so will the time required for adaptation increase towards that of T_1.

Summary

We are now in a position to summarize the answer, given by the intervening chapters, to the objection, raised in section 11/2, that ultrastability cannot be the mode of adaptation used by living organisms because it would take too long. We can now appreciate that the objection was unwittingly using the assumption that the organism and the environment were richly joined both within themselves and to each other. Evidence has been given, in section 15/2, that the *actual* richness is by no means high. Then Chapters 15 and 16 have shown that when it is not high, adaptation by ultrastability can occur in a time that is no longer impossibly long. Thus the objection has been answered, at least in outline.

There we must leave the matter, for a closer examination would have to depend on measurements of actual brains adapting to actual environments. The study of the matter should not be beyond the powers of the present-day experimenter.

[. . .]

12 F. E. Emery and E. L. Trist

The Causal Texture of Organizational Environments

F. E. Emery and E. L. Trist, 'The causal texture of organizational environments', *Human Relations*, vol. 18 (1965), pp. 21–32.

Identification of the Problem

A main problem in the study of organizational change is that the environmental contexts in which organizations exist are themselves changing, at an increasing rate, and towards increasing complexity. This point, in itself, scarcely needs labouring. Nevertheless, the characteristics of organizational environments demand consideration for their own sake, if there is to be an advancement of understanding in the behavioural sciences of a great deal that is taking place under the impact of technological change, especially at the present time. This paper is offered as a brief attempt to open up some of the problems, and stems from a belief that progress will be quicker if a certain extension can be made to current thinking about systems.

In a general way it may be said that to think in terms of systems seems the most appropriate conceptual response so far available when the phenomena under study – at any level and in any domain – display the character of being organized, and when understanding the nature of the interdependencies constitutes the research task. In the behavioural sciences, the first steps in building a systems theory were taken in connexion with the analysis of internal processes in organisms, or organizations, when the parts had to be related to the whole. Examples include the organismic biology of Jennings, Cannon, and Henderson; early Gestalt theory and its later derivatives such as balance theory; and the classical theories of social structure. Many of these problems could be represented in closed-system models. The next steps were taken when wholes had to be related to their environments. This led to open-system models.

A great deal of the thinking here has been influenced by

cybernetics and information theory, though this has been used as much to extend the scope of closed-system as to improve the sophistication of open-system formulations. It was von Bertalanffy (1950) who, in terms of the general transport equation which he introduced, first fully disclosed the importance of openness or closedness to the environment as a means of distinguishing living organisms from inanimate objects. In contradistinction to physical objects, any living entity survives by importing into itself certain types of material from its environment, transforming these in accordance with its own system characteristics, and exporting other types back into the environment. By this process the organism obtains the additional energy that renders it 'negentropic'; it becomes capable of attaining stability in a time-independent steady state – a necessary condition of adaptability to environmental variance.

Such steady states are very different affairs from the equilibrium states described in classical physics, which have far too often been taken as models for representing biological and social transactions. Equilibrium states follow the second law of thermodynamics, so that no work can be done when equilibrium is reached, whereas the openness to the environment of a steady state maintains the capacity of the organism for work, without which adaptability, and hence survival, would be impossible.

Many corollaries follow as regards the properties of open systems, such as equifinality, growth through internal elaboration, self-regulation, constancy of direction with change of position, etc. – and by no means all of these have yet been worked out. But though von Bertalanffy's formulation enables exchange processes between the organism, or organization, and elements in its environment to be dealt with in a new perspective, it does not deal at all with those processes in the environment itself which are among the determining conditions of the exchanges. To analyse these an additional concept is needed – *the causal texture of the environment* – if we may reintroduce, at a social level of analysis, a term suggested by Tolman and Brunswik (1935) and drawn from S. C. Pepper (1934).

With this addition, we may now state the following general proposition: that a comprehensive understanding of organizational behaviour requires some knowledge of each member of the following set, where L indicates some potentially lawful connexion, and the suffix 1 refers to the organization and the suffix 2 to the environment:

$$L_{1\,1}, L_{1\,2}$$
$$L_{2\,1}, L_{2\,2}$$

$L_{1\,1}$ here refers to processes within the organization – the area of internal interdependencies; $L_{1\,2}$ and $L_{2\,1}$ to exchanges between the organization and its environment – the area of transactional interdependencies, from either direction; and $L_{2\,2}$ to processes through which parts of the environment become related to each other – (i.e. its causal texture) – the area of interdependencies that belong within the environment itself.

In considering environmental interdependencies, the first point to which we wish to draw attention is that the laws connecting parts of the environment to each other are often incommensurate with those connecting parts of the organization to each other, or even with those which govern the exchanges. It is not possible, for example, always to reduce organization–environment relations to the form of 'being included in'; boundaries are also 'break' points. As Barker and Wright (1949), following Lewin (1936), have pointed out in their analysis of this problem as it affects psychological ecology, we may lawfully connect the actions of a javelin thrower in sighting and throwing his weapon; but we cannot describe in the same concepts the course of the javelin as this is affected by variables lawfully linked by meteorological and other systems.

The Development of Environmental Connectedness (Case 1)

A case history, taken from the industrial field, may serve to illustrate what is meant by the environment becoming organized at the social level. It will show how a greater degree of system-connectedness, of crucial relevance to the organization, may develop in the environment, which is yet not directly a function either of the organization's own characteristics or of its immediate relations. Both of these, of course, once again become crucial when the response of the organization to what has been happening is considered.

The company concerned was the foremost in its particular

market in the food-canning industry in the U.K. and belonged to a large parent group. Its main product – a canned vegetable – had some 65 per cent of this market, a situation which had been relatively stable since before the war. Believing it would continue to hold this position, the company persuaded the group board to invest several million pounds sterling in erecting a new, automated factory, which, however, based its economies on an inbuilt rigidity – it was set up exclusively for the long runs expected from the traditional market.

The character of the environment, however, began to change while the factory was being built. A number of small canning firms appeared, not dealing with this product nor indeed with others in the company's range, but with imported fruits. These firms arose because the last of the post-war controls had been removed from steel strip and tin, and cheaper cans could now be obtained in any numbers – while at the same time a larger market was developing in imported fruits. This trade being seasonal, the firms were anxious to find a way of using their machinery and retaining their labour in winter. They became able to do so through a curious side-effect of the development of quick-frozen foods, when the company's staple was produced by others in this form. The quick-freezing process demanded great constancy at the growing end. It was not possible to control this beyond a certain point, so that quite large crops unsuitable for quick freezing but suitable for canning became available – originally from another country (the United States) where a large market for quick-frozen foods had been established. These surplus crops had been sold at a very low price for animal feed. They were now imported by the small canners – at a better but still comparatively low price, and additional cheap supplies soon began to be procurable from underdeveloped countries.

Before the introduction of the quick-freezing form, the company's own canned product – whose raw material had been specially grown at additional cost – had been the premier brand, superior to other varieties and charged at a higher price. But its position in the product spectrum now changed. With the increasing affluence of the society, more people were able to afford the quick-frozen form. Moreover, there was competition from a great many other vegetable products which could substitute for the staple, and people preferred this greater variety. The advantage of being the premier line among canned forms diminished, and demand increased both for the not-so-expensive varieties among them and for the quick-frozen forms. At the same time, major changes were taking place in retailing; supermarkets were developing, and more and more large grocery chains were coming into existence. These establishments wanted to sell certain types of goods under their own house names, and began to place bulk orders with the small canners for their own varieties of the company's staple that fell within the class. As the small canners provided an extremely cheap article (having no marketing expenses and a cheaper raw material), they could undercut the manufacturers' branded product, and within three years they captured over 50 per cent of the market. Previously, retailers' varieties had accounted for less than 1 per cent.

The new automatic factory could not be adapted to the new situation until alternative products with a big sales volume could be developed, and the scale of research and development, based on the type of market analysis required to identify these, was beyond the scope of the existing resources of the company either in people or in funds.

The changed texture of the environment was not recognized by an able but traditional management until it was too late. They failed entirely to appreciate that a number of outside events were becoming connected with each other in a way that was leading up to irreversible general change. Their first reaction was to make an herculean effort to defend the traditional product, then the board split on whether or not to make entry into the cheaper unbranded market in a supplier role. Group H.Q. now felt they had no option but to step in, and many upheavals and changes in management took place until a 'redefinition of mission' was agreed, and slowly and painfully the company re-emerged with a very much altered product mix and something of a new identity.

Four Types of Causal Texture

It was this experience, and a number of others not dissimilar, by no means all of them industrial (and including studies of change problems in hospitals, in prisons, and in educational and political organizations), that gradually led us to feel a need for redirecting conceptual attention to the causal texture of the environment, considered as a quasi-independent domain. We have now isolated four 'ideal types' of causal texture, approximations to which may be thought of as existing simultaneously in the 'real world' of most organizations – though, of course, their weighting will vary enormously from case to case.

The first three of these types have already, and indeed repeatedly, been described – in a large variety of terms and with the emphasis on an equally bewildering variety of special aspects – in the literature of a number of disciplines, ranging from biology to economics and including military theory as well as psychology and sociology. The fourth type, however, is new, at least to us, and is the one that for some time we have been endeavouring to identify. About the first three, therefore, we can be brief, but the fourth is scarcely understandable without reference to them. Together, the four types may be said to form a series in which the degree of causal texturing is increased, in a new and significant way, as each step is taken. We leave as an open question the need for further steps.

Step one

The simplest type of environmental texture is that in which goals and noxiants ('goods' and 'bads') are relatively unchanging in themselves and randomly distributed. This may be called the *placid, randomized environment*. It corresponds to Simon's idea of a surface over which an organism can locomote: most of this is bare, but at isolated, widely scattered points there are little heaps of food (1957, p. 137). It also corresponds to Ashby's limiting case of no connexion between the environmental parts (1960, section 15/4); and to Schützenberger's random field (1954, p. 100). The economist's classical market also corresponds to this type.

A critical property of organizational response under random conditions has been stated by Schützenberger; that there is no distinction between tactics and strategy, 'the optimal strategy is just the simple tactic of attempting to do one's best on a purely local basis' (1954, p. 101). The best tactic, moreover, can be learnt only by trial and error and only for a particular class of local environmental variances (Ashby, 1960, p. 197). While organizations under these conditions can exist adaptively as single and indeed quite small units, this becomes progressively more difficult under the other types.

Step two

More complicated, but still a placid environment, is that which can be characterized in terms of clustering: goals and noxiants are not randomly distributed but hang together in certain ways. This may be called the *placid, clustered environment*, and is the case with which Tolman and Brunswik were concerned; it corresponds to Ashby's 'serial system' and to the economist's 'imperfect competition'. The clustering enables some parts to take on roles as signs of other parts or become means-objects with respect to approaching or avoiding. Survival, however, becomes precarious if an organization attempts to deal tactically with each environmental variance as it occurs.

The new feature of organizational response to this kind of environment is the emergence of strategy as distinct from tactics. Survival becomes critically linked with what an organization knows of its environment. To pursue a goal under its nose may lead it into parts of the field fraught with danger, while avoidance of an immediately difficult issue may lead it away from potentially rewarding areas. In the clustered environment the relevant objective is that of 'optimal location', some positions being discernible as potentially richer than others.

To reach these requires concentration of resources, subordination to the main plan, and the development of a 'distinctive competence', to use Selznick's (1957) term, in reaching the strategic objective. Organizations under these conditions, therefore, tend to grow in size and also to become hierarchical, with a tendency towards centralized control and co-ordination.

Step three

The next level of causal texturing we have called the *disturbed-reactive environment*. It may be compared with Ashby's ultrastable system or the economist's oligopolic market. It is a type 2 environment in which there is more than one organization of the same kind; indeed, the existence of a number of similar organizations now becomes the dominant characteristic of the environmental field. Each organization does not simply have to take account of the others when they meet at random, but has also to consider that what it knows can also be known by the others. The part of the environment to which it wishes to move itself in the long run is also the part to which the others seek to move. Knowing this, each will wish to improve its own chances by hindering the others, and each will know that the others must not only wish to do likewise, but also know that each knows this. The presence of similar others creates an imbrication, to use a term of Chein's (1943), of some of the causal strands in the environment.

If strategy is a matter of selecting the 'strategic objective' – where one wishes to be at a future time – and tactics a matter of selecting an immediate action from one's available repertoire, then there appears in type 3 environments to be an intermediate level of organizational response – that of the *operation* – to use the term adopted by German and Soviet military theorists, who formally distinguish tactics, operations, and strategy. One has now not only to make sequential choices, but to choose actions that will draw off the other organizations. The new element is that of deciding which of someone else's possible tactics one wishes to take place, while ensuring that others of them do not. An operation consists of a campaign involving a planned series of tactical initiatives, calculated reactions by others, and counter-actions. The flexibility required encourages a certain decentralization and also puts a premium on quality and speed of decision at various peripheral points (Heyworth, 1955).

It now becomes necessary to define the organizational objective in terms not so much of location as of capacity or power to move more or less at will, i.e. to be able to make and meet competitive challenge. This gives particular relevance to strategies of absorption and parasitism. It can also give rise to situations in which stability can be obtained only by a certain coming-to-terms between competitors, whether enterprises, interest groups, or governments. One has to know when not to fight to the death.

Step four

Yet more complex are the environments we have called *turbulent fields*. In these, dynamic processes, which create significant variances for the component organizations, arise from the field itself. Like type 3 and unlike the static types 1 and 2, they are dynamic. Unlike type 3, the dynamic properties arise not simply from the interaction of the component organizations, but also from the field itself. The 'ground' is in motion.

Three trends contribute to the emergence of these dynamic field forces:

i. The growth to meet type 3 conditions of organizations, and linked sets of organizations, so large that their actions are both persistent and strong enough to induce autochthonous processes in the environment. An analogous effect would be that of a company of soldiers marching in step over a bridge.
ii. The deepening interdependence between the economic and the other facets of the society. This means that economic organizations are increasingly enmeshed in legislation and public regulation.
iii. The increasing reliance on research and development to achieve the capacity to meet competitive challenge. This leads to a situation in which a change gradient is continuously present in the environmental field.

For organizations, these trends mean a gross increase in their area of *relevant uncertainty*. The consequences which flow from their actions lead off in ways that become increasingly unpredictable: they do not necessarily fall off with distance, but may at any point be amplified beyond all expectation; similarly, lines of action that are strongly pursued may find themselves attenuated by emergent field forces.

The Salience of Type 4 Characteristics (Case II)

Some of these effects are apparent in what happened to the canning company of case I, whose situation represents a transition from an environment largely composed of type 2 and type 3 characteristics to one where those of type 4 began to gain in salience. The case now to be presented illustrates the combined operation of the three trends described above in an altogether larger environmental field involving a total industry and its relations with the wider society.

The organization concerned is the National Farmers Union of Great Britain to which more than 200,000 of the 250,000 farmers of England and Wales belong. The present problem brought to us for investigation was that of communications. Headquarters felt, and was deemed to be, out of touch with county branches, and these with local branches. The farmer had looked to the N.F.U. very largely to protect him against market fluctuations by negotiating a comprehensive deal with the government at annual reviews concerned with the level of price support. These reviews had enabled home agriculture to maintain a steady state during two decades when the threat, or existence, of war in relation to the type of military technology then in being had made it imperative to maintain a high level of homegrown food without increasing prices to the consumer. This policy, however, was becoming obsolete as the conditions of thermonuclear stalemate established themselves. A level of support could no longer be counted upon which would keep in existence small and inefficient farmers – often on marginal land and dependent on family labour – compared with efficient medium-sized farms, to say nothing of large and highly mechanized undertakings.

Yet it was the former situation which had produced N.F.U. cohesion. As this situation receded, not only were farmers becoming exposed to more competition from each other, as well as from Commonwealth and European farmers, but the effects were being felt of very great changes which had been taking place on both the supply and marketing sides of the industry. On the supply side, a small number of giant firms now supplied almost all the requirements in fertilizer, machinery, seeds, veterinary products, etc. As efficient farming depended upon ever greater utilization of these resources, their controllers exerted correspondingly greater power over the farmers. Even more dramatic were the changes in the marketing of farm produce. Highly organized food processing and distributing industries had grown up dominated again by a few large firms, on contracts from which (fashioned to suit their rather than his interests) the farmer was becoming increasingly dependent. From both sides deep inroads were being made on his autonomy.

It became clear that the source of the felt difficulty about communications lay in radical environmental changes which were confronting the organization with problems it was ill-adapted to meet. Communications about these changes were being interpreted or acted upon as if they referred to the 'traditional' situation. Only through a parallel analysis of the environment and the N.F.U. was progress made towards developing understanding on the basis of which attempts to devise adaptive organizational policies and forms could be made. Not least among the problems was that of creating a bureaucratic élite that could cope with the highly technical long-range planning now required and yet remain loyal to the democratic values of the N.F.U. Equally difficult was that of developing mediating institutions – agencies that would effectively mediate the relations between agriculture and other economic sectors without triggering off massive competitive processes.

These environmental changes and the organizational crisis they induced were fully apparent two or three years before the question of Britain's possible entry into the Common Market first appeared on the political agenda – which, of course, further complicated every issue.

A workable solution needed to preserve reasonable autonomy for the farmers as an occupational group, while meeting the interests of other sections of the community. Any such possibility depended on securing the consent of the large majority of farmers to placing under some degree of N.F.U. control matters that hitherto had remained within their own power of decision. These included what they produced, how and to what standard, and how most of it should be marketed. Such thoughts were anathema, for however dependent the farmer had grown on the N.F.U. he also remained intensely individualistic. He was being asked, he now felt, to redefine his identity, reverse his basic values, and refashion his organization – all at the same time. It is scarcely surprising that progress has been, and remains, both fitful and slow, and ridden with conflict.

Values and Relevant Uncertainty

What becomes precarious under type 4 conditions is how organizational stability can be achieved. In these environments individual organizations, however large, cannot expect to adapt successfully simply through their own direct actions – as is evident in the case of the N.F.U. Nevertheless, there are some indications of a solution that may have the same general significance for these environments as have strategy and operations for types 2 and 3. This is the emergence of *values that have overriding significance for all members of the field.* Social values are here regarded as coping mechanisms that make it possible to deal with persisting areas of relevant uncertainty. Unable to trace out the consequences of their actions as these are amplified and resonated through their extended social fields, men in all societies have sought rules, sometimes categorical, such as the ten commandments, to provide them with a guide and ready calculus. Values are not strategies or tactics; as Lewin (1936) has pointed out, they have the conceptual character of 'power fields' and act as injunctions.

So far as effective values emerge, the character of richly joined, turbulent fields changes in a most striking fashion. The relevance of large classes of events no longer has to be sought in an intricate mesh of diverging causal strands, but is given directly in the ethical code. By this transformation a field is created which is no longer richly joined and turbulent but simplified and relatively static. Such a transformation will be regressive, or constructively adaptative, according to how far the emergent values adequately represent the new environmental requirements.

Ashby, as a biologist, has stated his view, on the one hand, that examples of environments that are both large and richly connected are not common, for our terrestrial environment is widely characterized by being highly subdivided (1960, p. 205); and, on the other, that, so far as they are encountered, they may well be beyond the limits of human adaptation, the brain being an ultrastable system. By contrast the role here attributed to social values suggests that this sort of environment may in fact be not only one to which adaptation is possible, however difficult, but one that has been increasingly characteristic of the human condition since the beginning of settled communities. Also, let us not forget that values can be rational as well as irrational and that the rationality of their rationale is likely to become more powerful as the scientific ethos takes greater hold in a society.

Matrix Organization and Institutional Success

Nevertheless, turbulent fields demand some overall form of organization that is essentially different from the hierarchically structured forms to which we are accustomed. Whereas type 3 environments require one or other form of accommodation between like, but competitive, organizations whose fates are to a degree negatively correlated, turbulent environments require some relationship between dissimilar organizations whose fates are, basically, positively correlated. This means relationships that will maximize co-operation and which recognize that no one organization can take over the role of 'the other' and become paramount. We are inclined to speak of this type of relationship as an *organizational matrix*. Such a matrix acts in the first place by delimiting on value criteria the character of what may be included in the field specified – and therefore who. This selectivity then enables some definable shape to be worked out without recourse to much in the way of formal hierarchy among members. Professional associations provide one model of which there has been long experience.

We do not suggest that in other fields than the professional the requisite sanctioning can be provided only by state-controlled bodies. Indeed, the reverse is far more likely. Nor do we suggest that organizational matrices will function so as to eliminate the need for other measures to achieve stability. As with values, matrix organizations, even if successful, will only help to transform turbulent environments into the kinds of environment we have discussed as 'clustered' and 'disturbed-reactive'. Though, with these transformations, an organization could hope to achieve a degree of stability through its strategies, operation, and tactics, the transformations would not provide environments identical with the originals. The strategic objective in the transformed cases could no longer be stated simply in terms of optimal location (as in type 2) or capabilities (as in type 3). It must now rather be formulated in terms of *institutionalization*. According to Selznick (1957) organizations become institutions through the embodiment of organizational values which relate them to the wider society.[1]

1. Since the present paper was presented, this line of thought has been further developed by Churchman and Emery (1964) in their discussion of

As Selznick has stated in his analysis of leadership in the modern American corporation, 'the default of leadership shows itself in an acute form when *organizational* achievement or survival is confounded with *institutional* success' (1957, p. 27). '. . . the executive becomes a statesman as he makes the transition from administrative management to institutional leadership' (1957, p. 154).

The processes of strategic planning now also become modified. In so far as institutionalization becomes a prerequisite for stability, the determination of policy will necessitate not only a bias towards goals that are congruent with the organization's own character, but also a selection of goal-paths that offer maximum convergence as regards the interests of other parties. This became a central issue for the N.F.U. and is becoming one now for an organization such as the National Economic Development Council, which has the task of creating a matrix in which the British economy can function at something better than the stop–go level.

Such organizations arise from the need to meet problems emanating from type 4 environments. Unless this is recognized, they will only too easily be construed in type 3 terms, and attempts will be made to secure for them a degree of monolithic power that will be resisted overtly in democratic societies and covertly in others. In the one case they may be prevented from ever undertaking their missions; in the other one may wonder how long they can succeed in maintaining them.

An organizational matrix implies what McGregor (1960) has called Theory Y. This in turn implies a new set of values. But values are psychosocial commodities that come into existence only rather slowly. Very little systematic work has yet been done on the establishment of new systems of values, or on the type of criteria that might be adduced to allow their effectiveness to be empirically tested. A pioneer attempt is that of Churchman and Ackoff (1950). Likert (1961) has suggested that, in the large

---

the relation of the statistical aggregate of individuals to structured role sets, 'Like other values, organizational values emerge to cope with relevant uncertainties and gain their authority from their reference to the requirements of larger systems within which people's interests are largely concordant.'

corporation or government establishment, it may well take some ten to fifteen years before the new type of group values with which he is concerned could permeate the total organization. For a new set to permeate a whole modern society the time required must be much longer – at least a generation, according to the common saying – and this, indeed, must be a minimum. One may ask if this is fast enough, given the rate at which type 4 environments are becoming salient. A compelling task for social scientists is to direct more research on to these problems.

Summary

1. A main problem in the study of organizational change is that the environmental contexts in which organizations exist are themselves changing – at an increasing rate, under the impact of technological change. This means that they demand consideration for their own sake. Towards this end a redefinition is offered, at a social level of analysis, of the causal texture of the environment, a concept introduced in 1935 by Tolman and Brunswik.

2. This requires an extension of systems theory. The first steps in systems theory were taken in connexion with the analysis of internal processes in organisms, or organizations, which involved relating parts to the whole. Most of these problems could be dealt with through closed-system models. The next steps were taken when wholes had to be related to their environments. This led to open-system models, such as that introduced by von Bertalanffy, involving a general transport equation. Though this enables exchange processes between the organism, or organization, and elements in its environment to be dealt with, it does not deal with those processes in the environment itself which are the determining conditions of the exchanges. To analyse these an additional concept – the causal texture of the environment – is needed.

3. The laws connecting parts of the environment to each other are often incommensurate with those connecting parts of the organization to each other, or even those which govern exchanges. Case history I illustrates this and shows the dangers and difficulties that arise when there is a rapid and gross increase in the area of relevant uncertainty, a characteristic feature of many contemporary environments.

4. Organizational environments differ in their causal texture, both as regards degree of uncertainty and in many other important respects. A typology is suggested which identifies four 'ideal types', approximations to which exist simultaneously in the 'real world' of most organizations, though the weighting varies enormously:

a. In the simplest type, goals and noxiants are relatively unchanging in themselves and randomly distributed. This may be called the placid, randomized environment. A critical property from the organization's viewpoint is that there is no difference between tactics and strategy, and organizations can exist adaptively as single, and indeed quite small, units.

b. The next type is also static, but goals and noxiants are not randomly distributed; they hang together in certain ways. This may be called the placid, clustered environment. Now the need arises for strategy as distinct from tactics. Under these conditions organizations grow in size, becoming multiple and tending towards centralized control and coordination.

c. The third type is dynamic rather than static. We call it the disturbed-reactive environment. It consists of a clustered environment in which there is more than one system of the same kind, i.e. the objects of one organization are the same as, or relevant to, others like it. Such competitors seek to improve their own chances by hindering each other, each knowing the others are playing the same game. Between strategy and tactics there emerges an intermediate type of organizational response – what military theorists refer to as operations. Control becomes more decentralized to allow these to be conducted. On the other hand, stability may require a certain coming-to-terms between competitors.

d. The fourth type is dynamic in a second respect, the dynamic properties arising not simply from the interaction of identifiable component systems but from the field itself (the 'ground'). We call these environments turbulent fields. The turbulence results from the complexity and multiple character of the causal interconnexions. Individual organizations, however large, cannot adapt successfully simply through their direct interactions. An examination is made of the enhanced importance of values, regarded as a basic response to persisting areas of relevant uncertainty, as providing a control mechanism, when commonly held by all members in a field. This raises the question of organizational forms based on the characteristics of a matrix.

5. Case history II is presented to illustrate problems of the transition from type 3 to type 4. The perspective of the four environmental types is used to clarify the role of Theory X and Theory Y as representing a trend in value change. The establishment of a new set of values is a slow social process requiring something like a generation – unless new means can be developed.

References

ASHBY, W. R. (1960), *Design for a Brain*, Chapman and Hall, rev. edn.
BARKER, R. G., and WRIGHT, H. F. (1949), 'Psychological ecology and the problem of psychosocial development', *Child Development*, vol. 20, pp. 131–43.
BERTALANFFY, L. VON (1950), 'The theory of open systems in physics and biology', *Science*, vol. 111, pp. 23–9.
CHEIN, I. (1943), 'Personality and typology', *Journal of Social Psychology*, vol. 18, pp. 89–101.
CHURCHMAN, C. W., and ACKOFF, R. L. (1950), *Methods of Inquiry*, Educational Publishers, St Louis.
CHURCHMAN, C. W., and EMERY, F. E. (1965), *Operational Research and the Social Sciences*, Tavistock Publications.
HEYWORTH, LORD (1955), *The Organization of Unilever*, Unilever Limited, London.
LEWIN, K. (1936), *Principles of Topological Psychology*, McGraw-Hill.
LEWIN, K. (1951), *Field Theory in Social Science*, Harper.
LIKERT, R. (1961), *New Patterns of Management*, McGraw-Hill.
MCGREGOR, D. (1960), *The Human Side of Enterprise*, McGraw-Hill.
PEPPER, S. C. (1934), 'The conceptual framework of Tolman's purposive behaviorism', *Psychological Review*, vol. 41, pp. 108–33.
SCHÜTZENBERGER, M. P. (1954), 'A tentative classification of goal-seeking behaviours', *Journal of Mental Science*, vol. 100, pp. 97–102.
SELZNICK, P. (1957), *Leadership in Administration*, Row-Peterson.
SIMON, H. A. (1957), *Models of Man*, Wiley.
TOLMAN, E. C., and BRUNSWIK, E. (1935), 'The organism and the causal texture of the environment', *Psychological Review*, vol. 42, pp. 43–77.

Part Four Human Organizations as Systems

The analysis of human organizations as systems is the common practice of most of the disciplines concerned with them. However, they typically confine themselves to such abstract analytical dimensions as economic allocation, information flows and allocation of power and responsibility. When the substantive organizations[1] are considered it is usual to find (a) a closed-systems approach, and (b) an either–or approach to the social and technical systems – the sociologists isolating the social system for consideration in itself and the systems engineers doing the same for the technical system.

There is no doubt that social systems analysis has advanced our knowledge of organizations much more rapidly than was being achieved by the earlier attention to part problems such as personnel selection and incentives. In view of the conceptual developments reflected in the preceding sections it does not seem possible that the social sciences can remain long at this level.

In the following readings we can sense some of the directions of development.

1. See Levy, M. J., *Structure of Society*, Princeton University Press, 1950, for a full discussion of this distinction between analytical and substantive.

13 P. Selznick

Foundations of the Theory of Organizations

P. Selznick, 'Foundations of the theory of organizations', *American Sociological Review*, vol. 13 (1948), pp. 25–35.

Trades unions, governments, business corporations, political parties, and the like are formal structures in the sense that they represent rationally ordered instruments for the achievement of stated goals. 'Organization,' we are told, 'is the arrangement of personnel for facilitating the accomplishment of some agreed purpose through the allocation of functions and responsibilities' (Gaus, 1936, p. 66). Or, defined more generally, formal organization is 'a system of consciously coordinated activities or forces of two or more persons' (Barnard, 1938, p. 73). Viewed in this light, formal organization is the structural expression of rational action. The mobilization of technical and managerial skills requires a pattern of coordination, a systematic ordering of positions and duties which defines a chain of command and makes possible the administrative integration of specialized functions. In this context *delegation* is the primordial organizational act, a precarious venture which requires the continuous elaboration of formal mechanisms of coordination and control. The security of all participants, and of the system as a whole, generates a persistent pressure for the institutionalization of relationships, which are thus removed from the uncertainties of individual fealty or sentiment. Moreover, it is necessary for the relations within the structure to be determined in such a way that individuals will be interchangeable and the organization will thus be free of dependence upon personal qualities (Parsons, 1937, p. 752). In this way, the formal structure becomes subject to calculable manipulation, an instrument of rational action.

But as we inspect these formal structures we begin to see that they never succeed in conquering the non-rational dimensions of organizational behavior. The latter remain at once indispensable to the continued existence of the system of coordination and at the same time the source of friction, dilemma, doubt, and ruin. This fundamental paradox arises from the fact that rational action systems are inescapably imbedded in an institutional matrix, in two significant senses: (1) the action system – or the formal structure of delegation and control which is its organizational expression – is itself only an aspect of a concrete social structure made up of individuals who may interact as *wholes*, not simply in terms of their formal roles within the system; (2) the formal system, and the social structure within which it finds concrete existence, are alike subject to the pressure of an institutional environment to which some over-all adjustment must be made. The formal administrative design can never adequately or fully reflect the concrete organization to which it refers, for the obvious reason that no abstract plan or pattern can – or may, if it is to be useful – exhaustively describe an empirical totality. At the same time, that which is not included in the abstract design (as reflected, for example, in a staff-and-line organization chart) is vitally relevant to the maintenance and development of the formal system itself.

Organization may be viewed from two standpoints which are analytically distinct but which are empirically united in a context of reciprocal consequences. On the one hand, any concrete organizational system is an *economy*; at the same time, it is an *adaptive social structure*. Considered as an economy, organization is a system of relationships which define the availability of scarce resources and which may be manipulated in terms of efficiency and effectiveness. It is the economic aspect of organization which commands the attention of management technicians and, for the most part, students of public as well as private administration (Gulick and Urwick, 1937; Urwick, 1943; Mooney and Reiley, 1939; and Dennison, 1931). Such problems as the span of executive control, the role of staff or auxiliary agencies, the relation of headquarters to field offices, and the relative merits of single or multiple executive boards are typical concerns of the science of administration. The coordinative scalar, and functional principles, as elements of the theory of organization, are products of the attempt to explicate the most general features of organization as a 'technical problem' or, in our terms, as an economy.

Organization as an economy is, however, necessarily conditioned by the organic states of the concrete structure, outside of the systematics of delegation and control. This becomes especially evident as the attention of leadership is directed toward such problems as the legitimacy of authority and the dynamics of persuasion. It is recognized implicitly in action and explicitly in the work of a number of students that the possibility of manipulating the system of coordination depends on the extent to which that system is operating within an environment of effective inducement to individual participants and of conditions in which the stability of authority is assured. This is in a sense the fundamental thesis of Barnard's remarkable study (1938). It is also the underlying hypothesis which makes it possible for Urwick to suggest that 'proper' or formal channels in fact function to 'confirm and record' decisions arrived at by more personal means (Urwick, 1943, p. 47). We meet it again in the concept of administration as a process of education, in which the winning of consent and support is conceived to be a basic function of leadership.[1] In short, it is recognized that control and consent cannot be divorced even within formally authoritarian structures.

The indivisibility of control and consent makes it necessary to view formal organizations as *cooperative* systems, widening the frame of reference of those concerned with the manipulation of organizational resources. At the point of action, of executive decision, the economic aspect of organization provides inadequate tools for control over the concrete structure. This idea may be readily grasped if attention is directed to the role of the individual within the organizational economy. From the standpoint of organization as a formal system, persons are viewed functionally, in respect to their *roles*, as participants in assigned segments of the cooperative system. But in fact individuals have a propensity to resist depersonalization, to spill over the boundaries of their segmentary roles, to participate as *wholes*. The formal systems (at an extreme, the disposition of 'rifles' at a military perimeter) cannot take account of the deviations thus introduced, and consequently break down as instruments of control when relied

1. See Gaus (1936). Studies of the problem of morale are instances of the same orientation, having received considerable impetus in recent years from the work of the Harvard Business School group.

upon alone. The whole individual raises new problems for the organization, partly because of the needs of his own personality, partly because he brings with him a set of established habits as well, perhaps, as commitments to special groups outside of the organization.

Unfortunately for the adequacy of formal systems of coordination, the needs of individuals do not permit a single-minded attention to the stated goals of the system within which they have been assigned. The hazard inherent in the act of delegation derives essentially from this fact. Delegation is an organizational act, having to do with formal assignments of functions and powers. Theoretically, these assignments are made to roles or official positions, not to individuals as such. In fact, however, delegation necessarily involves concrete individuals who have interests and goals which do not always coincide with the goals of the formal system. As a consequence, individual personalities may offer resistance to the demands made upon them by the official conditions of delegation. These resistances are not accounted for within the categories of coordination and delegation, so that when they occur they must be considered as unpredictable and accidental. Observations of this type of situation within formal structures are sufficiently commonplace. A familiar example is that of delegation to a subordinate who is also required to train his own replacement. The subordinate may resist this demand in order to maintain unique access to the 'mysteries' of the job, and thus insure his indispensability to the organization.

In large organizations, deviations from the formal system tend to become institutionalized, so that 'unwritten laws' and informal associations are established. Institutionalization removes such deviations from the realm of personality differences, transforming them into a persistent structural aspect of formal organizations.[2] These institutionalized rules and modes of informal cooperation are normally attempts by participants in the formal organization to control the group relations which form the environment of organizational decisions. The informal patterns (such as cliques)

2. The creation of informal structures within various types of organizations has received explicit recognition in recent years. See Roethlisberger and Dickson (1941), p. 524; Barnard (1938), chap. ix; and Moore (1946), chap. xv.

arise spontaneously, are based on personal relationships, and are usually directed to the control of some specific situation. They may be generated anywhere within a hierarchy, often with deleterious consequences for the formal goals of the organization, but they may also function to widen the available resources of executive control and thus contribute to rather than hinder the achievement of the stated objectives of the organization. The deviations tend to force a shift away from the purely formal system as the effective determinant of behavior to (1) a condition in which informal patterns buttress the formal, as through the manipulation of sentiment within the organization in favor of established authority; or (2) a condition wherein the informal controls effect a consistent modification of formal goals, as in the case of some bureaucratic patterns (Selznick, 1943). This trend will eventually result in the formalization of erstwhile informal activities, with the cycle of deviation and transformation beginning again on a new level.

The relevance of informal structures to organizational analysis underlines the significance of conceiving of formal organizations as cooperative systems. When the totality of interacting groups and individuals becomes the object of inquiry, the latter is not restricted by formal, legal, or procedural dimensions. The *state of the system* emerges as a significant point of analysis, as when an internal situation charged with conflict qualifies and informs actions ostensibly determined by formal relations and objectives. A proper understanding of the organizational process must make it possible to interpret changes in the formal system – new appointments or rules or reorganizations – in their relation to the informal and unavowed ties of friendship, class loyalty, power cliques, or external commitment. This is what it means 'to know the score'.

The fact that the involvement of individuals as whole personalities tends to limit the adequacy of formal systems of coordination does not mean that organizational characteristics are those of individuals. The organic, emergent character of the formal organization considered as a cooperative system must be recognized. This means that the *organization* reaches decisions, takes action, and makes adjustments. Such a view raises the question of the relation between organizations and persons. The significance of theoretical emphasis upon the cooperative *system* as such is derived from the insight that certain actions and consequences are enjoined independently of the personality of the individuals involved. Thus, if reference is made to the 'organization-paradox' – the tension created by the inhibitory consequences of certain types of informal structures within organizations – this does not mean that individuals themselves are in quandaries. It is the nature of the interacting consequences of divergent interests within the organization which creates the condition, a result which may obtain independently of the consciousness or the qualities of the individual participants. Similarly, it seems useful to insist that there are qualities and needs of leader*ship*, having to do with position and role, which are persistent despite variations in the character or personality of individual leaders themselves.

Rational action systems are characteristic of both individuals and organizations. The conscious attempt to mobilize available internal resources (e.g. self-discipline) for the achievement of a stated goal – referred to here as an economy or a formal system – is one aspect of individual psychology. But the personality considered as a dynamic system of interacting wishes, compulsions, and restraints defines a system which is at once essential and yet potentially deleterious to what may be thought of as the 'economy of learning' or to individual rational action. At the same time, the individual personality is an adaptive structure, and this, too, requires a broader frame of reference for analysis than the categories of rationality. On a different level, although analogously, we have pointed to the need to consider organizations as cooperative systems and adaptive structures in order to explain the context of and deviations from the formal systems of delegation and coordination.

To recognize the sociological relevance of formal structures is not, however, to have constructed a theory of organization. It is important to set the framework of analysis, and much is accomplished along this line when, for example, the nature of authority in formal organizations is reinterpreted to emphasize the factors of cohesion and persuasion as against legal or coercive sources (Michels, 1931; Barnard, 1938, chap. xii). This redefinition is logically the same as that which introduced the conception of the self as social. The latter helps make possible, but does not of itself fulfil, the requirements for a dynamic theory of personality. In the same way, the definition of authority as conditioned by sociological factors of sentiment and cohesion – or more generally the definition of formal organizations as cooperative systems – only sets the stage, as an initial requirement, for the formulation of a theory of organization.

Structural–Functional Analysis

Cooperative systems are constituted of individuals interacting as wholes in relation to a formal system of coordination. The concrete structure is therefore a resultant of the reciprocal influences of the formal and informal aspects of organization. Furthermore, this structure is itself a totality, an adaptive 'organism' reacting to influences upon it from an external environment. These considerations help to define the objects of inquiry; but to progress to a system of predicates *about* these objects it is necessary to set forth an analytical method which seems to be fruitful and significant. The method must have a relevance to empirical materials, which is to say, it must be more specific in its reference than discussions of the logic or methodology of social science.

The organon which may be suggested as peculiarly helpful in the analysis of adaptive structures has been referred to as 'structural–functional analysis' (Parsons, 1945). This method may be characterized in a sentence: *Structural–functional analysis relates contemporary and variable behavior to a presumptively stable system of needs and mechanisms.* This means that a given empirical system is deemed to have basic needs, essentially related to self-maintenance; the system develops repetitive means of self-defense; and day-to-day activity is interpreted in terms of the function served by that activity for the maintenance and defense of the system. Put thus generally, the approach is applicable on any level in which the determinate 'states' of empirically isolable systems undergo self-impelled and repetitive transformations when impinged upon by external conditions. This self-impulsion suggests the relevance of the term 'dynamic', which is often used in referring to physiological, psychological, or social systems to which this type of analysis has been applied.[3]

It is a postulate of the structural–functional approach that the basic need of all empirical systems is the maintenance of the integrity and continuity of the system itself. Of course, such a postulate is primarily useful in directing attention to a set of 'derived imperatives' or needs which are sufficiently concrete to characterize the system at hand.[4] It is perhaps rash to attempt a catalogue of these imperatives for formal organizations, but some suggestive formulation is needed in the interests of setting forth the type of analysis under discussion. In formal organizations, the 'maintenance of the system' as a generic need may be specified in terms of the following imperatives:

1. *The security of the organization as a whole in relation to social forces in its environment*. This imperative requires continuous attention to the possibilities of encroachment and to the forestalling of threatened aggressions or deleterious (though perhaps unintended) consequences from the actions of others.

2. *The stability of the lines of authority and communication.* One of the persistent reference points of administrative decision is the weighing of consequences for the continued capacity of leadership to control and to have access to the personnel or ranks.

3. *The stability of informal relations within the organization.* Ties of sentiment and self-interest are evolved as unacknowledged but effective mechanisms of adjustment of individuals and subgroups to the conditions of life within the organization. These ties represent a cementing of relationships which sustains the

3. 'Structure' refers to both the relationships within the system (formal plus informal patterns in organization) and the set of needs and modes of satisfaction which characterize the given type of empirical system. As the utilization of this type of analysis proceeds, the concept of 'need' will require further clarification. In particular, the imputation of a 'stable set of needs' to organizational systems must not function as a new instinct theory. At the same time, we cannot avoid using these inductions as to generic needs, for they help us to stake out our area of inquiry. The author is indebted to Robert K. Merton who has, in correspondence, raised some important objections to the use of the term 'need' in this context.

4. For 'derived imperative', see Malinowski (1945, pp. 44 ff.). For the use of 'need' in place of 'motive', see Malinowski (1944, pp. 89–90).

formal authority in day-to-day operations and widens opportunities for effective communication.[5] Consequently, attempts to 'upset' the informal structure, either frontally or as an indirect consequence of formal reorganization, will normally be met with considerable resistance.

4. *The continuity of policy and of the sources of its determination.* For each level within the organization, and for the organization as a whole, it is necessary that there be a sense that action taken in the light of a given policy will not be placed in continuous jeopardy. Arbitrary or unpredictable changes in policy undermine the significance of (and therefore the attention to) day-to-day action by injecting a note of capriciousness. At the same time, the organization will seek stable roots (or firm statutory authority or popular mandate) so that a sense of the permanency and legitimacy of its acts will be achieved.

5. *A homogeneity of outlook with respect to the meaning and role of the organization.* The minimization of disaffection requires a unity derived from a common understanding of what the character of the organization is meant to be. When this homogeneity breaks down, as in situations of internal conflict over basic issues, the continued existence of the organization is endangered. On the other hand, one of the signs of 'healthy' organization is the ability to effectively orient new members and readily slough off those who cannot be adapted to the established outlook.

This catalogue of needs cannot be thought of as final, but it approximates the stable system generally characteristic of formal organizations. These imperatives are derived, in the sense that they represent the conditions for survival or self-maintenance of cooperative systems of organized action. An inspection of these needs suggests that organizational survival is intimately connected with the struggle for relative prestige, both for the organization and for elements and individuals within it. It may therefore be useful to refer to a *prestige–survival motif* in organizational behavior as a short-hand way of relating behavior to needs, especially when the exact nature of the needs remains in doubt. However, it must be emphasized that prestige–survival in organizations does not derive simply from like motives in individuals.

5. They may also *destroy* those relationships, as noted above, but the need remains, generating one of the persistent dilemmas of leadership.

Loyalty and self-sacrifice may be individual expressions of organizational or group egotism and self-consciousness.

The concept of organizational need directs analysis to the *internal relevance* of organizational behavior. This is especially pertinent with respect to discretionary action undertaken by agents manifestly in pursuit of formal goals. The question then becomes one of relating the specific act of discretion to some presumptively stable organizational need. In other words, it is not simply action plainly oriented internally (such as in-service training) but also action presumably oriented externally which must be inspected for its relevance to internal conditions. This is of prime importance for the understanding of bureaucratic behavior, for it is of the essence of the latter that action formally undertaken for substantive goals be weighed and transformed in terms of its consequences for the position of the officialdom.

Formal organizations as cooperative systems on the one hand, and individual personalities on the other, involve structural–functional homologies, a point which may help to clarify the nature of this type of analysis. If we say that the individual has a stable set of needs, most generally the need for maintaining and defending the integrity of his personality or ego; that there are recognizable certain repetitive mechanisms which are utilized by the ego in its defense (rationalization, projection, regression, etc.); and that overt and variable behavior may be interpreted in terms of its relation to these needs and mechanisms – on the basis of this logic we may discern the typical pattern of structural–functional analysis as set forth above. In this sense, it is possible to speak of a 'Freudian model' for organizational analysis. This does not mean that the substantive insights of individual psychology may be applied to organizations, as in vulgar extrapolations from the individual ego to whole nations or (by a no-less-vulgar inversion) from strikes to frustrated workers. It is the *logic*, the *type* of analysis which is pertinent.

This homology is also instructive in relation to the applicability of generalizations to concrete cases. The dynamic theory of personality states a set of possible predicates about the ego and its mechanisms of defense, which inform us concerning the propensities of individual personalities under certain general circumstances. But these predicates provide only tools for the analysis of particular individuals, and each concrete case must be examined to tell which operate and in what degree. They are not primarily organs of prediction. In the same way, the predicates within the theory of organization will provide tools for the analysis of particular cases. Each organization, like each personality, represents a resultant of complex forces, an empirical entity which no single relation or no simple formula can explain. The problem of analysis becomes that of selecting among the possible predicates set forth in the theory of organization those which illuminate our understanding of the materials at hand.

The setting of structural–functional analysis as applied to organizations requires some qualification, however. Let us entertain the suggestion that the interesting problem in social science is not so much why men act the way they do as why men in certain circumstances *must* act the way they do. This emphasis upon constraint, if accepted, releases us from an ubiquitous attention to behavior in general, and especially from any undue fixation upon statistics. On the other hand, it has what would seem to be the salutary consequence of focusing inquiry upon certain necessary relationships of the type 'if . . . then', for example – if the cultural level of the rank and file members of a formally democratic organization is below that necessary for participation in the formulation of policy, then there will be pressure upon the leaders to use the tools of demagogy.

Is such a statement universal in its applicability? Surely not in the sense that one can predict without remainder the nature of all or even most political groups in a democracy. Concrete behavior is a resultant, a complex vector, shaped by the operation of a number of such general constraints. But there is a test of general applicability: it is that of noting whether the relation made explicit must be *taken into account* in action. This criterion represents an empirical test of the significance of social science generalizations. If a theory is significant it will state a relation which will either (a) be taken into account as an element of achieving control; or (b) be ignored only at the risk of losing control and will evidence itself in a ramification of objective or unintended consequences.[6] It is a corollary of this principle of

6. See R. M. MacIver's discussion of the 'dynamic assessment' which 'brings the external world selectively into the subjective realm, conferring

significance that investigation must search out the underlying factors in organizational action, which requires a kind of intensive analysis of the same order as psychoanalytic probing.

A frame of reference which invites attention to the constraints upon behavior will tend to highlight tensions and dilemmas, the characteristic paradoxes generated in the course of action. The dilemma may be said to be the handmaiden of structural–functional analysis, for it introduces the concept of *commitment* or *involvement* as fundamental to organizational analysis. A dilemma in human behavior is represented by an inescapable commitment which cannot be reconciled with the needs of the organism or the social system. There are many spurious dilemmas which have to do with verbal contradictions, but inherent dilemmas to which we refer are of a more profound sort, for they reflect the basic nature of the empirical system in question. An economic order committed to profit as its sustaining incentive may, in Marxist terms, sow the seed of its own destruction. Again, the anguish of man, torn between finitude and pride, is not a matter of arbitrary and replaceable assumptions but is a reflection of the psychological needs of the human organism, and is concretized in his commitment to the institutions which command his life; he is in the world and of it, inescapably involved in its goals and demands; at the same time, the needs of the spirit are compelling, proposing modes of salvation which have continuously disquieting consequences for worldly involvements. In still another context, the need of the human organism for affection and response necessitates a commitment to elements of the culture which can provide them; but the rule of the super-ego is uncertain since it cannot be completely reconciled with the need for libidinal satisfactions.

Applying this principle to organizations we may note that there is a general source of tension observable in the split between 'the motion and the act'. Plans and programs reflect the freedom of technical or ideal choice, but organized action cannot escape involvement, a commitment to personnel or institutions or pro-

---

on it subjective significance for the ends of action' (1942, chaps. 11 and 12). The analysis of this assessment within the context of organized action yields the implicit knowledge which guides the choice among alternatives. See also Merton (1936).

cedures which effectively qualifies the initial plan. *Der Mensch denkt, Gott lenkt.* In organized action, this ultimate wisdom finds a temporal meaning in the recalcitrance of the tools of action. We are inescapably committed to the mediation of human structures which are at once indispensable to our goals and at the same time stand between them and ourselves. The selection of agents generates immediately a bifurcation of interest, expressed in new centers of need and power, placing effective constraints upon the arena of action, and resulting in tensions which are never completely resolved. This is part of what it means to say that there is a 'logic' of action which impels us forward from one undesired position to another. Commitment to dynamic, self-activating tools is of the nature of organized action; at the same time, the need for continuity of authority, policy, and character are pressing, and require an unceasing effort to master the instruments generated in the course of action. This generic tension is specified within the terms of each cooperative system. But for all we find a persistent relationship between *need* and *commitment* in which the latter not only qualifies the former but unites with it to produce a continuous state of tension. In this way, the notion of constraint (as reflected in tension or paradox) at once widens and more closely specifies the frame of reference for organizational analysis.

For Malinowski, the core of functionalism was contained in the view that a cultural fact must be analysed in its setting. Moreover, he apparently conceived of his method as pertinent to the analysis of all aspects of cultural systems. But there is a more specific problem, one involving a principle of selection which serves to guide inquiry along significant lines. Freud conceived of the human organism as an adaptive structure, but he was not concerned with all human needs, nor with all phases of adaptation. For his system, he selected those needs whose expression is blocked in some way, so that such terms as repression, inhibition, and frustration became crucial. All conduct may be thought of as derived from need, and all adjustment represents the reduction of need. But not all needs are relevant to the systematics of dynamic psychology; and it is not adjustment as such but reaction to frustration which generates the characteristic modes of defensive behavior.

Organizational analysis, too, must find its selective principle; otherwise the indiscriminate attempts to relate activity functionally to needs will produce little in the way of significant theory. Such a principle might read as follows: *our frame of reference is to select out those needs which cannot be fulfilled within approved avenues of expression and thus must have recourse to such adaptive mechanisms as ideology and to the manipulation of formal processes and structures in terms of informal goals.* This formulation has many difficulties, and is not presented as conclusive, but it suggests the kind of principle which is likely to separate the quick and the dead, the meaningful and the trite, in the study of cooperative systems in organized action.[7]

The frame of reference outlined here for the theory of organization may now be identified as involving the following major ideas: (a) the concept of organizations as cooperative systems, adaptive social structures, made up of interacting individuals, sub-groups, and informal plus formal relationships; (b) structural–functional analysis, which relates variable aspects of organization (such as goals) to stable needs and self-defensive mechanisms; (c) the concept of recalcitrance as a quality of the tools of social action, involving a break in the continuum of adjustment and defining an environment of constraint, commitment, and tension. This frame of reference is suggested as providing a specifiable *area of relations* within which predicates in the theory of organization will be sought, and at the same time setting forth principles of selection and relevance in our approach to the data of organization.

It will be noted that we have set forth this frame of reference within the over-all context of social action. The significance of events may be defined by their place and operational role in a means–end scheme. If functional analysis searches out the elements important for the maintenance of a given structure, and that structure is one of the materials to be manipulated in action, then that which is functional in respect to the structure is also functional in respect to the action system. This provides a ground for the significance of functionally derived theories. At the same

7. This is not meant to deprecate the study of organizations as *economies* or formal systems. The latter represent an independent level, abstracted from organizational structures as cooperative or adaptive systems ('organisms').

time, relevance to control in action is the empirical test of their applicability or truth.

Cooptation as a Mechanism of Adjustment

The frame of reference stated above is in fact an amalgam of definition, resolution, and substantive theory. There is an element of *definition* in conceiving of formal organizations as cooperative systems, though of course the interaction of informal and formal patterns is a question of fact; in a sense, we are *resolving* to employ structural–functional analysis on the assumption that it will be fruitful to do so, though here, too, the specification of needs or derived imperatives is a matter for empirical inquiry; and our predication of recalcitrance as a quality of the tools of action is itself a *substantive theory*, perhaps fundamental to a general understanding of the nature of social action.

A theory of organization requires more than a general frame of reference, though the latter is indispensable to inform the approach of inquiry to any given set of materials. What is necessary is the construction of generalizations concerning transformations within and among cooperative systems. These generalizations represent, from the standpoint of particular cases, possible predicates which are relevant to the materials as we know them in general, but which are not necessarily controlling in all circumstances. A theory of transformations in organization would specify those states of the system which resulted typically in predictable, or at least understandable, changes in such aspects of organization as goals, leadership, doctrine, efficiency, effectiveness, and size. These empirical generalizations would be systematized as they were related to the stable needs of the cooperative system.

Changes in the characteristics of organizations may occur as a result of many different conditions, not always or necessarily related to the processes of organization as such. But the theory of organization must be selective, so that explanations of transformations will be sought within its own assumptions or frame of reference. Consider the question of size. Organizations may expand for many reasons – the availability of markets, legislative delegations, the swing of opinion – which may be accidental from the point of view of the organizational process. To explore changes in size (as of, say, a trades union) as related to changes in non-organizational conditions may be necessitated by the historical events to be described, but it will not of itself advance the frontiers of the theory of organization. However, if 'the innate propensity of all organizations to expand' is asserted as a function of 'the inherent instability of incentives' (Barnard, 1938, pp. 158–9) then transformations have been stated within the terms of the theory of organization itself. It is likely that in many cases the generalization in question may represent only a minor aspect of the empirical changes, but these organizational relations must be made explicit if the theory is to receive development.

In a frame of reference which specifies needs and anticipates the formulation of a set of self-defensive responses or mechanisms, the latter appear to constitute one kind of empirical generalization or 'possible predicate' within the general theory. The needs of organizations (whatever investigation may determine them to be) are posited as attributes of all organizations, but the responses to disequilibrium will be varied. The mechanisms used by the system in fulfilment of its needs will be repetitive and thus may be described as a specifiable set of assertions within the theory of organization, but any given organization may or may not have recourse to the characteristic modes of response. Certainly no given organization will employ all of the possible mechanisms which are theoretically available. When Barnard speaks of an 'innate propensity of organization to expand' he is in fact formulating one of the general mechanisms, namely, expansion, which is a characteristic mode of response available to an organization under pressure from within. These responses necessarily involve a transformation (in this case, size) of some structural aspect of the organization.

Other examples of the self-defensive mechanisms available to organizations may derive primarily from the response of these organizations to the institutional environments in which they live. The tendency to construct ideologies, reflecting the need to come to terms with major social forces, is one such mechanism. Less well understood as a mechanism of organizational adjustment is what we may term *cooptation*. Some statement of the meaning of this concept may aid in clarifying the foregoing analysis.

Cooptation is the process of absorbing new elements into the leadership or policy-determining structure of an organization as a means of averting threats to its stability or existence. This is a defensive mechanism, formulated as one of a number of possible predicates available for the interpretation of organizational behavior. Cooptation tells us something about the process by which an institutional environment impinges itself upon an organization and effects changes in its leadership and policy. Formal authority may resort to cooptation under the following general conditions:

1. When there exists a hiatus between consent and control, so that the legitimacy of the formal authority is called into question. The 'indivisibility' of consent and control refers, of course, to an optimum situation. Where control lacks an adequate measure of consent, it may revert to coercive measures or attempt somehow to win the consent of the governed. One means of winning consent is to coopt elements into the leadership or organization, usually elements which in some way reflect the sentiment, or possess the confidence of the relevant public or mass. As a result, it is expected that the new elements will lend respectability or legitimacy to the organs of control and thus reestablish the stability of formal authority. This process is widely used, and in many different contexts. It is met in colonial countries, where the organs of alien control reaffirm their legitimacy by coopting native leaders into the colonial administration. We find it in the phenomenon of 'crisis-patriotism' wherein normally disfranchised groups are temporarily given representation in the councils of government in order to win their solidarity in a time of national stress. Cooptation is presently being considered by the United States Army in its study of proposals to give enlisted personnel representation in the court-martial machinery – a clearly adaptive response to stresses made explicit during the war, the lack of confidence in the administration of army justice. The 'unity' parties of totalitarian states are another form of cooptation; company unions or some employee representation plans in industry are still another. In each of these cases, the response of formal authority (private or public, in a large organization or a small one) is an attempt to correct a state of imbalance by *formal* measures. It will be noted, moreover, that what is shared is the

responsibility for power rather than power itself. These conditions define what we shall refer to as *formed cooptation.*

2. Cooptation may be a response to the pressure of specific centers of power. This is not necessarily a matter of legitimacy or of a general and diffuse lack of confidence. These may be well established; and yet organized forces which are able to threaten the formal authority may effectively shape its structure and policy. The organization in respect to its institutional environment – or the leadership in respect to its ranks – must take these forces into account. As a consequence, the outside elements may be brought into the leadership or policy-determining structure, may be given a place as a recognition of and concession to the resources they can independently command. The representation of interests through administrative constituencies is a typical example of this process. Or, within an organization, individuals upon whom the group is dependent for funds or other resources may insist upon and receive a share in the determination of policy. This form of cooptation is typically expressed in informal terms, for the problem is not one of responding to a state of imbalance with respect to the 'people as a whole' but rather one of meeting the pressure of specific individuals or interest-groups which are in a position to enforce demands. The latter are interested in the substance of power and not its forms. Moreover, an open acknowledgement of capitulation to specific interests may itself undermine the sense of legitimacy of the formal authority within the community. Consequently, there is a positive pressure to refrain from explicit recognition of the relationship established. This form of the cooptative mechanism, having to do with the sharing of power as a response to specific pressures, may be termed *informal cooptation*

Cooptation reflects a state of tension between formal authority and social power. The former is embodied in a particular structure and leadership, but the latter has to do with subjective and objective factors which control the loyalties and potential manipulability of the community. Where the formal authority is an expression of social power, its stability is assured. On the other hand, when it becomes divorced from the sources of social power its continued existence is threatened. This threat may arise from the sheer alienation of sentiment or from the fact that other leaderships have control over the sources of social power. Where a formal authority has been accustomed to the assumption that its constituents respond to it as individuals, there may be a rude awakening when organization of those constituents on a non-governmental basis creates nuclei of power which are able effectively to demand a sharing of power.[8]

The significance of cooptation for organizational analysis is not simply that there is a change in or a broadening of leadership, and that this is an adaptive response, but also that *this change is consequential for the character and role of the organization.* Cooptation involves commitment, so that the groups to which adaptation has been made constrain the field of choice available to the organization or leadership in question. The character of the coopted elements will necessarily shape (inhibit or broaden) the modes of action available to the leadership which has won adaptation and security at the price of commitment. The concept of cooptation thus implicitly sets forth the major points of the frame of reference outlined above: it is an adaptive response of a cooperative system to a stable need, generating transformations which reflect constraints enforced by the recalcitrant tools of action.

References

BARNARD, C. I. (1938), *The Functions of the Executive*, Harvard University Press.

DENNISON, H. S. (1931), *Organization Engineering*, McGraw-Hill.

GAUS, J. M. (1936), 'A theory of organization in public administration', *The Frontiers of Public Administration*, University of Chicago Press.

GULICK, L., and URWICK, L. (1937), *Papers on the Science of Administration*, Institute of Public Administration, Columbia University.

8. It is perhaps useful to restrict the concept of cooptation to formal organizations, but in fact it probably reflects a process characteristic of all group leaderships. This has received some recognition in the analysis of class structure, wherein the ruling class is interpreted as protecting its own stability by absorbing new elements. Thus Michels made the point that 'an aristocracy cannot maintain an enduring stability by sealing itself off hermetically'. See Michels (1934, p. 39) and Mosca (1939, pp. 413 ff.). The alliance or amalgamation of classes in the face of a common threat may be reflected in formal and informal cooptative responses among formal organizations sensitive to class pressure. In Selznick (1949) the author has made extensive use of the concept of cooptation in analysing some aspects of the organizational behavior of a government agency.

MacIver, R. M. (1942), *Social Causation*, Ginn.

Malinowski, B. (1944), *A Scientific Theory of Culture*, University of North Carolina Press.

Malinowski, B. (1945), *The Dynamics of Culture Change*, Yale University Press.

Merton, R. K. (1936), 'The unanticipated consequences of purposive social action', *American Sociological Review*, vol. 1, no. 6.

Michels, R. (1931), 'Authority', in *Encyclopedia of the Social Sciences*, Macmillan, pp. 319 ff.

Michels, R. (1934), *Umschichtungen in den herrschenden Klassen nach dem Kriege*, Kohlhammer.

Mooney, J. D., and Reiley, A. C. (1939), *The Principles of Organization*, Harper.

Moore, W. E. (1946), *Industrial Relations and the Social Order*, Macmillan.

Mosca, G. (1939), *The Ruling Class*, McGraw-Hill.

Parsons, T. (1937), *The Structure of Social Action*, McGraw-Hill.

Parsons, T. (1945), 'The present position and prospects of systematic theory in sociology', in G. Gurvitch and W. E. Moore (eds.), *Twentieth-Century Sociology*, The Philosophical Library, New York.

Roethlisberger, F. J., and Dickson, W. J. (1941), *Management and the Worker*, Harvard University Press.

Selznick, P. (1943), 'An approach to a theory of bureaucracy', *American Sociological Review*, vol. 8, no. 1.

Selznick, P. (1949), *TVA and the Grass Roots*, University of California Press.

Urwick, L. (1943), *The Elements of Administration*, Harper.

14 F. E. Emery and E. L. Trist

Socio-technical Systems

F. E. Emery and E. L. Trist, 'Socio-technical systems', in C. W. Churchman and M. Verhulst (eds.) *Management Science, Models and Techniques*, vol. 2, Pergamon, 1960, pp. 83–97.

The analysis of the characteristics of enterprises as systems would appear to have strategic significance for furthering our understanding of a great number of specific industrial problems. The more we know about these systems the more we are able to identify what is relevant to a particular problem and to detect problems that tend to be missed by the conventional framework of problem analysis.

The value of studying enterprises as systems has been demonstrated in the empirical studies of Blau (4), Gouldner (6), Jaques (8), Selznick (15) and Lloyd Warner (21). Many of these studies have been informed by a broadly conceived concept of bureaucracy, derived from Weber and influenced by Parsons and Merton:

> They have found their main business to be in the analysis of a specific bureaucracy as a complex social system, concerned less with the individual differences of the actors than with the situationally shaped roles they perform (6).

Granted the importance of system analysis there remains the important question of whether an enterprise should be construed as a 'closed' or an 'open system', i.e. relatively 'closed' or 'open' with respect to its external environment. Von Bertalanffy (3) first introduced this general distinction in contrasting biological and physical phenomena. In the realm of social theory, however, there has been something of a tendency to continue thinking in terms of a 'closed' system, that is, to regard the enterprise as sufficiently independent to allow most of its problems to be analysed with reference to its internal structure and without reference to its external environment. Early exceptions were Rice and Trist (11) in the field of labour turnover and Herbst (7) in the analysis of social flow systems. As a first step, closed system thinking has been fruitful, in psychology and industrial sociology, in directing attention to the existence of structural similarities, relational determination and subordination of part to whole. However, it has tended to be misleading on problems of growth and the conditions for maintaining a 'steady state'. The formal physical models of 'closed systems' postulate that, as in the second law of thermodynamics, the inherent tendency of such systems is to grow towards maximum homogeneity of the parts and that a steady state can only be achieved by the cessation of all activity. In practice, the system theorists in social science (and these include such key anthropologists as Radcliffe-Brown) refused to recognize these implications but instead, by the same token, did '*tend* to focus on the statics of social structure and to neglect the study of structural change' (10). In an attempt to overcome this bias, Merton suggested that 'the concept of dysfunction, which implies the concept of strain, stress and tension on the structural level, provides an analytical approach to the study of dynamics and change' (10). This concept has been widely accepted by system theorists but while it draws attention to sources of imbalance within an organization it does not conceptually reflect the mutual permeation of an organization and its environment that is the cause of such imbalance. It still retains the limiting perspectives of 'closed system' theorizing. In the administrative field the same limitations may be seen in the otherwise invaluable contributions of Barnard (2) and related writers.

The alternative conception of 'open systems' carries the logical implications that such systems may spontaneously reorganize towards states of greater heterogeneity and complexity and that they achieve a 'steady state' at a level where they can still do work. Enterprises appear to possess at least these characteristics of 'open systems'. They grow by processes of internal elaboration (7) and manage to achieve a steady state while doing work, i.e. achieve a quasi-stationary equilibrium in which the enterprise as a whole remains constant, with a continuous '*throughput*', despite a considerable range of external changes (9, 11).

The appropriateness of the concept of 'open system' can be settled, however, only by examining in some detail what is involved in an enterprise achieving a steady state. The continued existence of any enterprise presupposes some regular commerce in products or services with other enterprises, institutions and persons in its external social environment. If it is going to be useful to speak of steady states in an enterprise, they must be states in which this commerce is going on. The conditions for regularizing this commerce lie both within and without the enterprise. On the one hand, this presupposes that an enterprise has at its immediate disposal the necessary material supports for its activities – a workplace, materials, tools and machines – and a work force able and willing to make the necessary modifications in the material 'throughput' or provide the requisite services. It must also be able, efficiently, to utilize its material supports and to organize the actions of its human agents in a rational and predictable manner. On the other hand, the regularity of commerce with the environment may be influenced by a broad range of independent external changes affecting markets for products and inputs of labour, materials and technology. If we examine the factors influencing the ability of an enterprise to maintain a steady state in the face of these broader environmental influences we find that:

1. The variation in the output markets that can be tolerated without structural change is a function of the flexibility of the technical productive apparatus – its ability to vary its rate, its end product or the mixture of its products. Variation in the output markets may itself be considerably reduced by the display of distinctive competence. Thus the output markets will be more attached to a given enterprise if it has, relative to other producers, a distinctive competence – a distinctive ability to deliver the right product to the right place at the right time.
2. The tolerable variation in the 'input' markets is likewise dependent upon the technological component. Thus some enterprises are enabled by their particular technical organization to tolerate considerable variation in the type and amount of labour they can recruit. Others can tolerate little.

The two significant features of this state of affairs are:

1. That there is no simple one-to-one relation between variations in inputs and outputs. Depending upon the technological system, different combinations of inputs may be handled to yield similar outputs and different 'product mixes' may be produced from similar inputs. As far as possible an enterprise will tend to do these things rather than make structural changes in its organization. It is one of the additional

characteristics of 'open systems' that while they are in constant commerce with the environment they are selective and, within limits, self-regulating.

2. That the technological component, in converting inputs into outputs, plays a major role in determining the self-regulating properties of an enterprise. It functions as one of the major boundary conditions of the social system of the enterprise in thus mediating between the ends of an enterprise and the external environment. Because of this the materials, machines and territory that go to making up the technological component are usually defined, in any modern society, as 'belonging' to an enterprise and excluded from similar control by other enterprises. They represent, as it were, an 'internalized environment'.

Thus the mediating boundary conditions must be represented among 'the open system constants' (3) that define the conditions under which a steady state can be achieved. The technological component has been found to play a key mediating role and hence it follows that the open system concept must be referred to the socio-technical system, not simply to the social system of an enterprise.

It might be justifiable to exclude the technological component from the system concept if it were true, as many writers imply, that it plays only a passive and intermittent role. However, it cannot be dismissed as simply a set of limits that exert an influence at the initial stage of building an enterprise and only at such subsequent times as these limits are overstepped. There is, on the contrary, an almost constant accommodation of stresses arising from changes in the external environment; the technological component not only sets limits upon what can be done, but also in the process of accommodation creates demands that must be reflected in the internal organization and ends of an enterprise.

Study of a productive system therefore requires detailed attention to both the technological and the social components. It is not possible to understand these systems in terms of some arbitrarily selected single aspect of the technology such as the repetitive nature of the work, the coerciveness of the assembly conveyor or the piecemeal nature of the task. However, this is what is usually attempted by students of the enterprise. In fact:

It has been fashionable of late, particularly in the 'human relations' school, to assume that the actual job, its technology, and its mechanical and physical requirements are relatively unimportant compared to the social and psychological situation of men at work (5).

Even when there has been a detailed study of the technology this has not been systematically related to the social system but been treated as background information (21).

In our earliest study of production systems in coal mining it became apparent that 'so close is the relationship between the various aspects that the social and the psychological can be understood only in terms of the detailed engineering facts and of the way the technological system as a whole behaves in the environment of the underground situation' (19).

An analysis of a technological system in these terms can produce a systematic picture of the tasks and task inter-relations required by a technological system. However, between these requirements and the social system there is not a strictly determined one-to-one relation but what is logically referred to as a correlative relation.

In a very simple operation such as manually moving and stacking railway sleepers ('ties') there may well be only a single suitable work relationship structure, namely, a co-operating pair with each man taking an end of the sleeper and lifting, supporting, walking and throwing in close co-ordination with the other man. The ordinary production process is much more complex and there it is unusual to find that only one particular work relationship structure can be fitted to these tasks.

This element of choice and the mutual influence of technology and the social system may both be illustrated from our studies, made over several years, of work organization in British deep-seam coal mining. The following data are adapted from Trist and Murray (20).

Thus Table 1 indicates the main features of two very different forms of organization that have both been operated economically within the same seam and with identical technology.

The conventional system combines a complex formal structure with simple work roles: the composite system combines a simple formal structure with complex work roles. In the former the miner has a commitment to only a single part task and enters into only a very limited number of unvarying social relations that are sharply divided between those within his particular task group and those who are outside. With those 'outside' he shares no sense of belongingness and he recognizes no responsibility to them for the consequences of his actions. In the composite system the miner has a commitment to the whole group task and consequently finds himself drawn into a variety of tasks in co-operation with different members of the total group; he may be drawn into any task on the coal-face with any member of the total group.

Table 1
Same Technology, Same Coal Seam, Different Social Systems

	A conventional cutting longwall mining system	*A composite cutting longwall mining system*
Number of men	41	41
Number of completely segregated task groups	14	1
Mean job variation for members:		
task groups worked with	1·0	5·5
main tasks worked	1·0	3·6
different shifts worked	2·0	2·9

That two such contrasting social systems can effectively operate the same technology is clear enough evidence that there exists an element of choice in designing a work organization.

However, it is not a matter of indifference which form of organization is selected. As has already been stated, the technological system sets certain requirements of its social system and the effectiveness of the total production system will depend upon the adequacy with which the social system is able to cope with these requirements. Although alternative social systems may survive in that they are both accepted as 'good enough' (17) this does not preclude the possibility that they may differ in effectiveness.

In this case the composite systems consistently showed a superiority over the conventional in terms of production and costs.

This superiority reflects, in the first instance, the more adequate coping in the composite system with the task requirements. The constantly changing underground conditions require that the already complex sequence of mining tasks undergo frequent changes in the relative magnitudes and even the order of these tasks. These conditions optimally require the internal flexibility possessed in

Table 2
Production and Costs for Different Forms of Work Organization with Same Technology

	'Conventional'	*'Composite'*
Productive achievement*	78	95
Ancillary work at face (hours per man-shift)	1·32	0·03
Average reinforcement of labour (per cent of total face force)	6	—
Per cent of shifts with cycle lag	69	5
Number consecutive weeks without losing a cycle	12	65

* Average per cent of coal won from each daily cut, corrected for differences in seam transport.

varying degrees by the composite systems. It is difficult to meet variable task requirements with any organization built on a rigid division of labour. The only justification for a rigid division of labour is a technology which demands specialized non-substitute skills and which is, moreover, sufficiently superior, as a technology, to offset the losses due to rigidity. The conventional longwall cutting system has no such technical superiority over the composite to offset its relative rigidity – its characteristic inability to cope with changing conditions other than by increasing the stress placed on its members, sacrificing smooth cycle progress or drawing heavily upon the negligible labour reserves of the pit.

The superiority of the composite system does not rest alone in more adequate coping with the tasks. It also makes better provision to the personal requirements of the miners. Mutually supportive relations between task groups are the exception in the conventional system and the rule in the composite. In consequence, the conventional miner more frequently finds himself without support from his fellows when the strain or size of his task requires it. Crises are more likely to set him against his fellows and hence worsen the situation.

Similarly, the distribution of rewards and statuses in the conventional system reflects the relative bargaining power of different roles and task groups as much as any true differences in skill and effort. Under these conditions of disparity between effort and reward any demands for increased effort are likely to create undue stress.

The following table indicates the differences in stress experienced by miners in the two systems.

Table 3
Stress Indices for Different Social Systems

	'Conventional'	*'Composite'*
Absenteeism (per cent of possible shifts)		
without reason	4·3	0·4
sickness or other	8·9	4·6
accidents	6·8	3·2
Total	20·0	8·2

These findings were replicated by experimental studies in textile mills in the radically different setting of Ahmedabad, India (12).

However, two possible sources of misunderstanding need to be considered:

1. Our findings do not suggest that work group autonomy should be maximized in all productive settings. There is an optimum level of grouping which can be determined only by analysis of the requirements of the technological system. Neither does there appear to be any simple relation between level of mechanization and level of grouping. In one mining study we found that in moving from a hand-filling to a machine-filling technology, the appropriate organization shifted from an undifferentiated composite system to one based on a number of partially

segregated task groups with more stable differences in internal statuses.
2. Nor does it appear that the basic psychological needs being met by grouping are workers' needs for friendship on the job, as is frequently postulated by advocates of better 'human relations' in industry. Grouping produces its main psychological effects when it leads to a system of work roles such that the workers are primarily related to each other by way of the requirements of task performance and task interdependence. When this task orientation is established the worker should find that he has an adequate range of mutually supportive roles (mutually supportive with respect to performance and to carrying stress that arises from the task). As the role system becomes more mature and integrated, it becomes easier for a worker to understand and appreciate his relation to the group. Thus in the comparison of different composite mining groups it was found that the differences in productivity and in coping with stress were not primarily related to differences in the level of friendship in the groups. The critical prerequisites for a composite system are an adequate supply of the required special skills among members of the group and conditions for developing an appropriate system of roles. Where these prerequisites have not been fully met, the composite system has broken down or established itself at a less than optimum level. The development of friendship and particularly of mutual respect occurs in the composite systems but the friendship tends to be limited by the requirements of the system and not assume unlimited disruptive forms such as were observed in conventional systems and were reported by Adams (1) to occur in certain types of bomber crews.

The textile studies (12) yielded the additional finding that *supervisory roles* are best designed on the basis of the same type of socio-technical analysis. It is not enough simply to allocate to the supervisor a list of responsibilities for specific tasks and perhaps insist upon a particular style of handling men. The supervisory roles arise from the need to control and co-ordinate an incomplete system of men–task relations. Supervisory responsibility for the specific parts of such a system is not easily reconcilable with responsibility for overall aspects. The supervisor who continually intervenes to do some part of the productive work may be proving his willingness to work but is also likely to be neglecting his main task of controlling and co-ordinating the system so that the operators are able to get on with their jobs with the least possible disturbance.

Definition of a supervisory role presupposes analysis of the system's requirements for control and co-ordination and provision of conditions that will enable the supervisor readily to perceive what is needed of him and to take appropriate measures. As his control will in large measure rest on his control of the boundary conditions of the system – those activities relating to a larger system

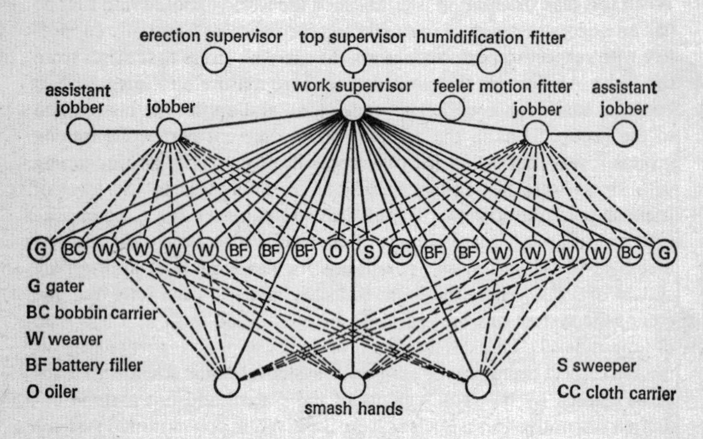

Figure 1 Management hierarchy before change

– it will be desirable to create 'unified commands' so that the boundary conditions will be correspondingly easy to detect and manage. If the unified commands correspond to natural task groupings, it will also be possible to maximize the autonomous responsibility of the work group for internal control and co-ordination, thus freeing the supervisor for his primary task. A graphic illustration of the differences in a supervisory role following a socio-technical reorganization of an automatic loom shed (12) can be seen in the following two figures. Figure 1 representing the situation before and Figure 2 representing the situation after change.

This reorganization was reflected in a significant and sustained improvement in mean percentage efficiency and a decrease in mean percentage damage.

The significance of the difference between these two organiza-

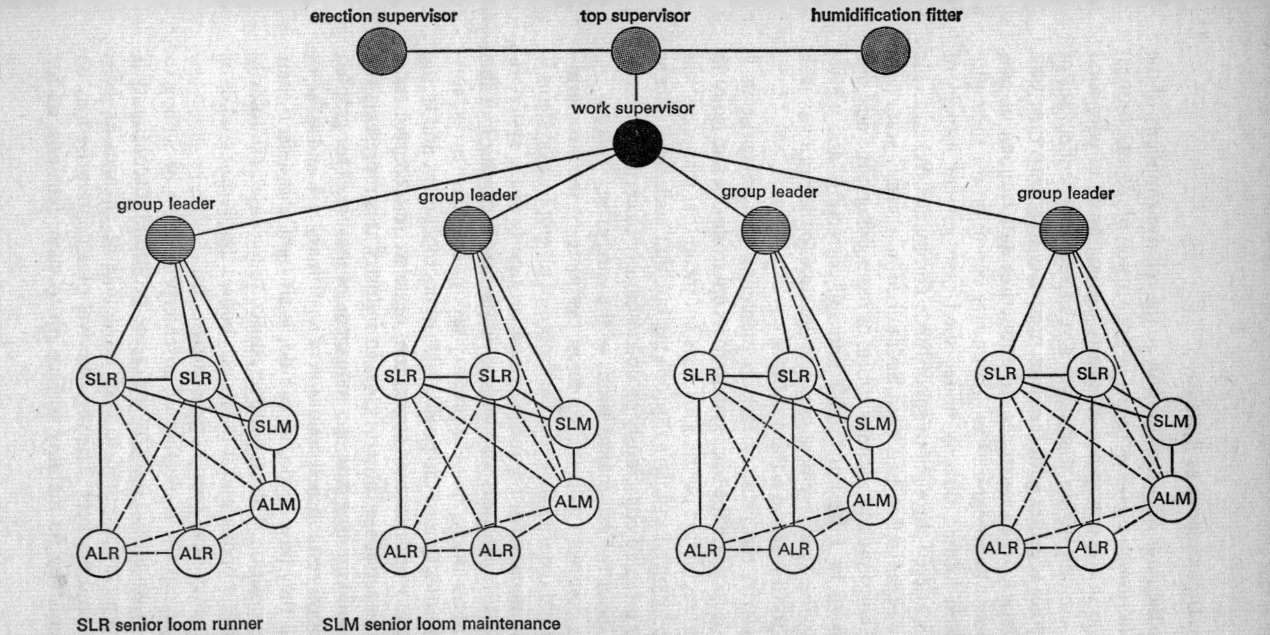

Figure 2 Management hierarchy after change

tional diagrams does not rest only in the relative simplicity of the latter (although this does reflect less confusion of responsibilities) but also in the emergence of clearly distinct areas of command which contain within themselves a relatively independent set of work roles together with the skills necessary to govern their task boundaries. In like manner the induction and training of new members was recognized as a boundary condition for the entire shed and located directly under shed management instead of being scattered throughout subordinate commands. Whereas the former organization had been maintained in a steady state only by the constant and arduous efforts of management, the new one proved to be inherently stable and self-correcting, and consequently freed management to give more time to their primary task and also to manage a third shift.

Similarly, the primary task in managing the enterprise as a whole is to relate the total system to its environment and is not in internal regulation *per se*. This does not mean that managers will not be involved in internal problems but that such involvement will be oriented consciously or unconsciously to certain assumptions about the external relations of the enterprise.

This contrasts with the common postulate of the structural-functional theories that 'the basic need of all empirical systems is the maintenance of the integrity and continuity of the system itself' (14). It contrasts also with an important implication of this postulate, namely, that the primary task of management is 'continuous attention to the possibilities of encroachment and to the forestalling of threatened aggressions or deleterious consequences from the actions of others' (14). In industry this represents the special and limiting case of a management that takes for granted a previously established definition of its primary task and assumes that all they have to do, or can do, is sit tight and defend their market position. This is, however, the common case in statutorily established bodies and it is on such bodies that recent studies of bureaucracy have been largely carried out.

In general the leadership of an enterprise must be willing to break down an old integrity or create profound discontinuity if such steps are required to take advantage of changes in technology and markets. The very survival of an enterprise may be threatened by its inability to face up to such demands, as for instance, switching the main effort from production of processed goods to marketing or from production of heavy industrial goods to consumer goods. Similarly, the leadership may need to pay 'continuous' attention to the possibilities of making their own encroachments rather than be obsessed with the possible encroachments of others.

Considering enterprises as 'open socio-technical systems' helps to provide a more realistic picture of how they are both influenced by and able to act back on their environment. It points in particular to the various ways in which enterprises are enabled by their structural and functional characteristics ('system constants') to cope with the 'lacks' and 'gluts' in their available environment. Unlike mechanical and other inanimate systems they possess the property of 'equifinality'; they may achieve a steady state from differing initial conditions and in differing ways (3). Thus in coping by internal changes they are not limited to simple quantitative change and increased uniformity but may, and usually do, elaborate new structures and take on new functions. The cumulative effect of coping mainly by *internal* elaboration and differentiation is generally to make the system independent of an increasing range of the predictable fluctuations in its supplies and outlets. At the same time, however, this process ties down in specific ways more and more of its capital, skill and energies and renders it less able to cope with newly emergent and unpredicted changes that challenge the primary ends of the enterprise. This process has been traced out in a great many empirical studies of bureaucracies (4, 10, 15).

However, there are available to an enterprise other aggressive strategies that seek to achieve a steady state by transforming the environment. Thus an enterprise has some possibilities for moving into new markets or inducing changes in the old, for choosing differently from among the range of personnel, resources and technologies offered by its environment or training and making new ones, and for developing new consumer needs or stimulating old ones.

Thus, arising from the nature of the enterprise as an open system, management is concerned with 'managing' both an internal system and an external environment. To regard an enterprise as a closed system and concentrate upon management of the 'internal enterprise' would be to expose the enterprise to the full impact of the vagaries of the environment.

If management is to control internal growth and development it must in the first instance control the 'boundary conditions' – the forms of exchange between the enterprise and its environment. As we have seen most enterprises are confronted with a multitude of actual and possible exchanges. If resources are not to be dissipated the management must select from the alternatives a course of action. The casual texture of competitive environments is such that it is extremely difficult to survive on a simple strategy of selecting the best from among the alternatives immediately offering. Some that offer immediate gain lead nowhere, others lead to greater loss; some alternatives that offer loss are avoidable, others are unavoidable if long-run gains are to be made. The relative size of the immediate loss or gain is no sure guide as to what follows. Since also the actions of an enterprise can improve the alternatives that are presented to it, the optimum course is more likely to rest in selecting a strategic objective to be achieved in the long run. The strategic objective should be to place the enterprise in a position in its environment where it has some assured conditions for growth – unlike war the best position is not necessarily that of unchallenged monopoly. Achieving this position would be the *primary task* or overriding mission of the enterprise.

In selecting the primary task of an enterprise, it needs to be borne in mind that the relations with the environment may vary with: (a) the productive efforts of the enterprise in meeting environmental requirements, (b) changes in the environment that may be induced by the enterprise and (c) changes independently taking place in the environment. These will be of differing importance for different enterprises and for the same enterprises at different times. Managerial control will usually be greatest if the primary task can be based on productive activity. If this is not possible, as in commerce, the primary task will give more control if it is based on marketing than simply on foreknowledge of the independent environmental changes. Managerial control will be further enhanced if the primary task, at whatever level it is selected, is such as to enable the enterprise to achieve *vis-à-vis* its competitors, a *distinctive competence*. Conversely, in our experience, an enterprise which has long occupied a favoured position because of distinctive productive competence may have grave difficulty in recognizing when it is losing control owing to environmental changes beyond its control.

As Selznick has pointed out (16), an appropriately defined primary task offers stability and direction to an enterprise, protecting it from adventurism or costly drifting. These advantages, however, as he illustrates (16), may be no more than potential unless the top management group of the organization achieves solidarity about the new primary task. If the vision of the task is locked up in a single man or is the subject of dissension in top management it will be subject to great risk of distortion and susceptible to violent fluctuations. Similarly, the enterprise as a whole needs to be reoriented and reintegrated about this primary task. Thus, if the primary task shifts from heavy industrial goods to durable consumer goods it would be necessary to ensure that there is a corresponding shift in values that are embodied in such sections as the sales force and design department.

References

1. S. Adams, 'Status congruency as a variable in small group performance', *Social Forces*, vol. 32 (1953), pp. 16–22.
2. C. I. Barnard, *The Functions of the Executive*, Harvard University Press, 1948.
3. L. von Bertalanffy, 'The theory of open systems in physics and biology', *Science*, vol. 111 (1950), pp. 23–9.
4. P. Blau, *The Dynamics of Bureaucratic Structure: A Study of Interpersonal Relations in Two Government Agencies*, University of Chicago Press, 1955.
5. P. F. Drucker, 'The Employee Society', *American Sociological Review*, vol. 58 (1952), pp. 358–63.
6. A. W. Gouldner, *Patterns of Industrial Bureaucracy*, Routledge and Kegan Paul, 1955.
7. P. G. Herbst, 'The analysis of social flow systems', *Human Relations*, vol. 7 (1954), pp. 327–36.
8. E. Jaques, *The Changing Culture of a Factory*, Tavistock, 1951.
9. K. Lewin, *Field Theory in Social Science*, Harper, 1951.
10. R. K. Merton, *Social Theory and Social Structure*, Free Press, 1949.
11. A. K. Rice and E. L. Trist, 'Institutional and sub-institutional determinants of change in labour turnover (The Glacier Project – VIII)', *Human Relations*, vol. 5 (1952), pp. 347–72.
12. A. K. Rice, *Productivity and Social Organization: The Ahmedabad Experiment*, Tavistock, 1958.
13. M. P. Schützenberger, 'A tentative classification of goal-seeking behaviours', *Journal of Mental Science*, vol. 100 (1954) pp. 97–102.

14. P. SELZNICK, 'Foundations of the theory of organization', *American Sociological Review*, vol. 13 (1948), pp. 25–35.
15. P. SELZNICK, *TVA and the Grass Roots*, University of California Press, 1949.
16. P. SELZNICK, *Leadership in Administration*, Row-Peterson, 1957.
17. H. A. SIMON, *Models of Man*, Wiley, 1957.
18. E. C. TOLMAN and E. BRUNSWIK, 'The organism and the causal texture of the environment', *Psychological Review*, vol. 42 (1935), pp. 43–77.
19. E. L. TRIST and K. W. BAMFORTH, 'Some social and psychological consequences of the longwall method of coal-getting', *Human Relations*, vol. 4 (1951).
20. E. L. TRIST and H. MURRAY, 'Work organization at the coal face: a comparative study of mining systems', *Tavistock Institute of Human Relations, Doc.*, no. 506 (1948).
21. W. L. WARNER and J. O. LOW, *The Social System of the Modern Factory*, Yale University, Press, 1947.

15 E. Nagel

A Formalization of Functionalism

E. Nagel, 'A formalization of functionalism', *Logic Without Metaphysics*, Free Press, 1956, pp. 247–83.

For some years now outstanding students in the social sciences have been urging and debating the use of so-called 'functional analysis' in their disciplines. In the opening chapter of his book, Professor Merton (1949) has surveyed sympathetically but critically much of the literature of functionalism, noted some of the ambiguities and other unclarities in the formulations of its nature and objectives, and offered what he calls a tentative 'paradigm of the concepts and objectives' of functional analysis with the aim of supplying a heuristically more adequate codification of this approach in sociology. There would be little profit in going over the literature that he has canvassed in such masterly fashion. However, functionalism in the social sciences has admittedly been inspired, and continues to be influenced, by the supposed character of functional analyses in physiology. Merton acknowledges this inspiration and influence; but his paradigm nevertheless does not make explicit how its various parts are related to elements in the functional approach in biology. It is the aim of the present essay to examine Merton's paradigm in the light of a schema of distinctions derived from an analysis of functional explanations in biological science. This examination is undertaken with one primary objective in mind: to exhibit the several items in Merton's codification as intimately related features in a coherent pattern of analysis, and thereby to make more evident than he has done the indispensable requirements which an adequate functional account in sociology must seek to satisfy.

I

We shall first present in outline the main features of functional analyses in physiology, and will begin with some general remarks preliminary to the main discussion.

1. Functional statements in biology (sometimes also called 'teleological statements') can be readily identified by the occurrence in them of certain characteristic locutions. For example, 'the function of the liver in animal bodies is to store sugar', 'plants have roots in order to enable them to absorb liquids in the soil', 'amoeba have cilia for the purpose of locomotion', and 'flowers secrete nectar so as to attract insects which will pollinate the ovaries', are typical functional statements. They contain such distinctive expressions as 'the function of', 'in order that', 'so as to', 'for the purpose of', and several others of similar nature. Expressions of this kind are rarely if ever used in modern science in relation to the behavior of inanimate bodies that have not been deliberately constructed to achieve certain effects (end results or 'goals'). But they are frequently employed in statements about inanimate physical systems which have been devised with some end in view, as in the statements 'the function of a clock is to tell time' and 'the role of the governor in a stationary engine is to regulate the opening of the throttle and thereby the speed of the rotating shaft'.

(a) *Some senses of the word 'function'*. It will be evident that the sense of the word 'function' as used in the preceding paragraph does not coincide with the meaning of the word as it is employed in mathematical analysis and its various special applications. In mathematics a function is a class of ordered pairs of elements (y, x), where those members of the pairs which are values of the (so-called 'independent') variable x are called the 'arguments', while those which are values of the (so-called 'dependent') variable y are the values 'corresponding to' the xs. To say that one variable is a function of another in this sense, is to assert that there is a rule which determines the values of the dependent variable from the values of the independent variable in a given range. But to say that the function of some organ in a living body (or of some part in a machine) is such-and-such, is to assert that the organ and some of its activities (and correspondingly for a machine part) is instrumental to maintaining some state or process of the organism, so that the occurrence of this state or process is causally dependent upon that organ and its behavior. The familiar general characterization of the task of science as that of establishing 'functional correlations' between variables is usually not intended to mean that all the sciences engage in functional analyses in the manner of physiology.

However, biologists themselves appear to use the word 'function' in several ways. (i) They sometimes employ it in such contexts as 'the function of an organ' to refer to the role it plays in some system of which it is a part. This corresponds to the sense of the word illustrated earlier, and it is a sense which will occupy us in the sequel. (ii) They also employ it in such contexts as 'the vital functions', to refer to certain broad types of organic processes (reproduction, assimilation, respiration, etc.) which are either definitory of a living organism, or are indispensable to the continuance of the organism or the species. In this context, the word 'function' signifies not the role of an organ (or sets of organs) in some system, but a general consequence or end product (or a set of consequences) of the operations of otherwise unspecified organic parts. (iii) But biologists also employ the word in such contexts as 'the functioning of an organ' (e.g. 'the functioning of the stomach') to refer to some or all of the processes occurring in that organ (e.g. muscular contractions, secretion of acids, absorption of liquids, etc.) where the function in sense (i) of these processes may or may not be specified. It is essential for the sake of clarity to keep in mind these different senses of the word 'function', especially since the failure to do so uniformly on the part of functional analysis in sociology has been the source of much confusion (Woodger, 1929).

One further source of possible misunderstanding deserves brief mention. Few if any contemporary biologists impute conscious aims or intentions to the organs (and other parts of living organisms) whose functions they investigate. Thus, the cilia of amoeba are not assumed by most current investigators to have aims or purposes, even if locomotion is often said to be 'the purpose' of cilia. Accordingly, despite the frequent use of such expressions as 'for the purpose of' or 'in order that' in the formulation of functional analyses in biology, this use must not be construed as implying a belief in the operation in organic processes of conscious goals or deliberate ends in view. In short, the roles which organic parts are discovered to have in a system are not to be understood as if they were causal agents that generate the behavior of those parts.

(b) *Functional and nonfunctional formulations.* Although functional statements in physiology are recognizable by the distinctive locutions they employ, their factual content can nevertheless be exhaustively rendered by nonfunctional formulations that are more customary in other parts of natural science. Consider the functional statement 'the function of the kidneys is to maintain a steady chemical composition of the blood.' It expresses the effect (or one effect) of the kidneys' structure and activities upon the chemistry of the blood. And this fact can also be expressed without the use of functional language by stating the conditions (or a condition) under which the indicated effect is obtained, in the following manner: 'in living organisms of a certain type, a necessary condition for a steady chemical composition of the blood is the presence and activity of the kidneys', or 'the blood in a living organism of a certain kind continues to exhibit a steady chemical composition only if the kidneys behave in a specified manner'. The functional statement appears to be equivalent in content to either of these nonfunctional ones. The sole difference between them is that, while in the former we are concerned to state the consequences for a specified system of the activities of one of its parts, in the latter we stress the condition under which a certain trait or state of the system is manifested. The difference between a functional and a nonfunctional formulation is thus one of selective emphasis; it is quite comparable to the difference between saying that B is the effect of A, and saying that A is the condition (or cause) of B.

But despite this equivalence, it is notorious that in many disciplines (e.g. modern physics) functional statements rarely occur. No modern student of electricity, for example, would feel comfortable with a proposed reformulation of Ohm's Law (at constant temperature, the current in a wire is proportional to the electro-motive force) into a functional statement such as 'at constant temperature, the current in a wire varies in order to remain proportional to the electro-motive force'. Several reasons can be advanced to account for this sense of discomfort, but perhaps the most important one is the following. There is a prima facie difference between most systems investigated in the physical sciences and those studied in biology. In the former, the properties and activities of the systems are dependent upon a set of factors in such a way that when these factors undergo any considerable variation the properties and activities of the system cease to exist. In the latter, however, the systems appear to be self-maintaining with respect to the continued manifestation of certain of their traits, despite fairly extensive variations in the factors upon which those traits causally depend. For example, the temperature of a stone will fluctuate with the temperature of its environment. But because the human body possesses mechanisms of regulation, it can maintain a fairly constant internal temperature despite quite considerable variations in the temperature of its environment. Accordingly, functional statements are regarded as appropriate in connection with systems possessing self-maintaining mechanisms for certain of their traits, but seem pointless and even misleading when used with reference to systems lacking such self-regulatory devices (Nagel, 1953).

We now turn to a closer inspection of the general character of such 'directively organized' (sometimes also called 'goal-directed') systems.

2. Let us first briefly consider an actual example of a 'directively organized' (or 'functional') system in biology, which we shall then use to elicit a generalized formulation of what appears to be distinctive of such systems.

Walter B. Cannon has shown with rich detail that the human body is capable of maintaining many of its traits in a fairly constant (or steady) state (homeostasis), as a consequence of possessing various coordinated physiological mechanisms (Cannon, 1932). One of the most striking of these traits is the internal temperature of the adult body, which is maintained between fairly narrow limits despite extensive changes in the temperature of the environment, and which cannot much exceed these limits without fatal results to the human organism. Elementary physical considerations thus make it clear that the body must possess a mechanism that can compensate for temperature variations of the environment, provided that these latter are not too extreme, and thereby enables the body to perform its customary activities in relative independence of the environmental temperature changes. Among the cyclic physiological processes that regulate the internal temperature are the following: if the body temperature is raised, the thyroid gland becomes less active and the basal metabolism is lowered, and vice versa; the quantity of blood flowing through peripheral blood vessels, and therefore the heat conducted or radiated through the skin, is altered by the dilation and contraction of these vessels; the amount of adrenalin secreted into the blood, and thereby the internal combustion of the body, is influenced by temperature changes; the moisture evaporated from the skin also affects the internal temperature, and the amount of moisture secreted depends on the respiration rate and the activity of the sweat glands which in turn are sensitive to temperature changes: and the automatic muscular contractions that are manifested as shivering provide a further source of internal heat.

(a) We shall now generalize from this example, and make explicit the elements that are involved in any directively organized system.[1]

Let *S* be such a system and *E* its 'external' environment. Just how the line between *S* and *E* is to be drawn need not concern us; it is a problem which must be settled on the basis of the special facts in individual cases, though it is conceivable that in some cases the line is drawn quite arbitrarily. We are supposing that *S* is a 'functional' (or 'self-maintaining', or 'directively organized', or 'goal-directed') system with respect to a certain trait (property, state, process) *G*. That is, *S* either possesses *G* at some time or during some period, or *S* is undergoing a series of alterations terminating in *G*, such that *S* is preserved in the state *G* or in its development to acquire *G*, despite some fairly extensive class of changes whether in *E* or in certain parts of *S* itself. We are therefore assuming that unless *S* contains some mechanism which produces effects compensating for these changes, it will cease to exhibit *G* or the tendency to acquire *G*; and it will be our primary task to make this assumption more articulate.

1. The following discussion is heavily indebted to Sommerhoff (1950), and in lesser measure to Ashby (1952).

It is of utmost importance to specify in each concrete case both the system S and the trait G. For in the first place, a system may be self-maintaining with respect to one trait but not with respect to some other. Thus, the human organism exhibits homeostasis with respect to its internal temperature, but apparently not with respect to the diameter of the iris of the eye. In the second place, S may be a part of some more inclusive system S', and though S is directively organized relative to G, S' may not be. In the third place, there may be several Gs with respect to all of which S is a functional system. Nevertheless, as will become clearer in the sequel, the circumstances under which S is self-maintaining with respect to some of these Gs may not coincide with the circumstances under which it is self-maintaining with respect to the others. Moreover, some of the Gs with respect to which S is self-maintaining may constitute some kind of 'hierarchy' – the hierarchy might be based on relations of causal dependence, temporal precedence, relations of inclusiveness or specificity, importance in some scale of values, and so on – and the conditions under which S is self-maintaining with respect to one member of the hierarchy may or may not be compatible with the self-maintenance of S with respect to another member of the hierarchy. Each of these possibilities will be found to have considerable relevance for the subsequent discussion.

(b) To continue the analysis we require the notions of 'state coordinate' (or 'state variable') and 'state description' which play such important roles in theoretical physics; and we must stop to explain them.

Imagine a physical system Σ to be completely isolated from external influences and to exhibit at some initial time t_0 specific forms of the set of properties Γ (which may or may not be exhaustive of all the properties of Σ), which will be denoted by 'Γ_0'. If left undisturbed until time t_1, the system will exhibit the same or different specific forms of these properties Γ, which will be denoted by 'Γ_1'. Suppose now that Σ is brought back into its initial state Γ_0, and that after an interval of time $(t_1 - t_0)$ it once again exhibits Γ_1. If Σ behaves in an analogous way, no matter what state is taken as the initial state and no matter how large the temporal interval, it will be called a 'deterministic system with respect to Γ' (or more briefly, when there is no danger of confusion, a 'deterministic system'). The set of properties Γ may be quite extensive, perhaps far too numerous for convenient observation. However, assume that there are n (a relatively small finite number) properties in Γ, whose specific forms can be taken as values of the variables $x_1, x_2, \ldots, x_n$, such that the specific forms of *all* the properties in Γ at any time are uniquely determined by these n properties at that time, and such that n is the smallest number of properties for which this holds. Accordingly, if at some initial time t_0 these variables have the values $x_1^{t_0}, \ldots, x_n^{t_0}$ (so that Σ is in the state Γ_0), while at some subsequent time t_1 the variables have the values $x_1^{t_1}, \ldots, x_n^{t_1}$ (so that Σ is in the state Γ_1), then since Σ is a deterministic system the second set of values of the variables (and hence the second state of Σ) are uniquely determined by the first set. Such variables will be called 'state coordinates', and the set of variables will be referred to as a 'state description'.

It is explicitly postulated that the values of the state coordinates at a given time are independent of one another, even though their values at one time depend on their values at another time. (However, it is not excluded that there may be more than one set of state coordinates for a system. In this case, though the 'instantaneous' values of the coordinates in each set will be independent of one another, the values of the coordinates in one set will not be independent of the values of the coordinates in the other set at a given time.) The relations of dependence between the values of state coordinates at different times (i.e. the 'dynamical laws' of the system) may not be known. But if the state coordinates are numerical variables (so that the properties they represent can be measured), these relations can be represented schematically as follows:

$$x_1^t = f_1(x_1^{t_0}, \ldots, x_n^{t_0}; t - t_0)$$
$$\cdots\cdots\cdots\cdots$$
$$x_n^t = f_n(x_1^{t_0}, \ldots, x_n^{t_0}, t - t_0)$$

where the f_is are single-valued functions, and the first derivates of the x_i with respect to the time are also single-valued functions whose arguments are these variables and no other functions of the time. This last proviso can be stated thus:

$$\frac{dx_1}{dt} = F_1 (x_1, \ldots, x_n)$$

$$\cdot \quad \cdot \quad \cdot \quad \cdot \quad \cdot \quad \cdot \quad \cdot$$

$$\frac{dx_n}{dt} = F_n (x_1, \ldots, x_n),$$

where the F_i are single-valued but not necessarily continuous.

However, it is not essential to the subsequent discussion that the state coordinates be numerical variables, so that we are free to employ the ideas here presented even in connection with systems whose properties are not strictly measurable.

An illustration from elementary physics will give point to this abstract account. Mechanics investigates various properties of bodies associated with their motions (such as relative positions, velocities, accelerations, moments of inertia, kinetic energy, and the like), and it is frequently possible (as in astronomy) to approximate to the ideal of a completely isolated mechanical system. In the mechanics of point particles (i.e. in the study of bodies whose dimensions can be neglected), the state coordinates are the position and momentum of each particle in the system under study. Thus, in the case of a freely falling body, it suffices to know (in addition, of course, to the laws of motion) the position and momentum of the body at some initial instant, in order to be able to calculate its position and momentum (and accordingly other properties of the body, such as its kinetic energy, which are definable in terms of these coordinates) at any other instant. It should also be noted that the position of a particle at any given time is independent of its momentum at that time, even though its position at one time depends on its position and momentum at some other time.

(c) We can now resume the discussion of systems S which are self-maintaining with respect to some G. Assume for the present that the environment E remains constant, so that it can be ignored; and suppose that G is causally dependent on the presence and operation of n related parts into which S is analysable. Let us further stipulate for the sake of simplicity, but without loss of generality, that there are just three parts of S that are thus causally relevant to G; and we shall employ as state coordinates for these parts the variables A, B and C respectively, with numerical subscripts to indicate the specific form of the complex properties that characterize these parts at any given time. Accordingly, the state of S at any time that is causally relevant to G can be indicated by a specialization of the matrix $(A_xB_yC_z)$. Thus, to illustrate these conventions in terms of a familiar example, if S is the human organism, G is its internal temperature falling into a certain interval, and G depends only on the condition of three parts of S, the state of S at any given time that is causally relevant to G can be obtained as a specialization of the matrix $(A_xB_yC_z)$, where A_x is the degree of dilation of the peripheral blood vessels, B_y the degree of activity of the thyroid glands, and C_z the respiration rate.

(d) Thus far, no restrictions have been placed upon the possible values of the state coordinates. But some restrictions must in fact be imposed, if the analysis is to single out the distinctive features of functional systems. Thus, in the example just given, it would be absurd to suppose that the peripheral blood vessels could be distended so that their mean diameter is five feet (and that therefore A_x could have five feet as one of its values). Accordingly, we shall stipulate as the *first* condition upon the possible values of the state coordinates A_x, B_y and C_z, that these values fall into certain classes K_A, K_B and K_C respectively, where the width of the classes will in general depend both on the special character of the parts of S represented by the corresponding state coordinates as well as on the special nature of the system S. We can also formulate this restriction as the requirement that the possible states of S which are specializations of the matrix $(A_xB_yC_z)$, must all fall into a certain class K_S, whose width will depend on the special facts already mentioned.

(e) Now suppose that during some period of time T and at some initial time t_0 in this period, S is in the state $(A_0B_0C_0)$, such that S either exhibits the property G or will acquire G if left undisturbed. Call such a state of S a 'causally effective state with respect to G', or a 'G state' for short. It is clear that not every state of S which satisfies the assumptions thus far made need be a G state; for although every state must be a member of K_S, it may happen that one of the state coordinates has a value such that for no permissible values of the remaining coordinates at that time will *S* be in a *G* state. Thus, if in the example of the human body as the system *S* with *G* as an internal temperature of from 97° to 99° F, should B_y (the coordinate for the condition of the thyroid gland) have some extreme value at a given time – corresponding, say, to acute hyperactivity – no permissible values of the remaining state coordinates at that time may suffice to yield a *G* state for *S*. In short, not every combination of possible values of the state coordinates is a *G* state for *S*; and the occurrence of a *G* state will depend on certain special requirements being satisfied, over and above those involved in the determination of the class K_R.

Let the class of *G* states of *S* which are determined by these special requirements be K_G. Clearly, K_G must be a subclass of K_S. However, K_G may be empty, so that no possible state of *S* is a *G* state. Thus, if *G* is an internal temperature of from 150° to 160° F, then *G* is apparently never realizable in the system *S* consisting of the human body. On the other hand, K_G may contain more than one member, though at a given time only one of them can be realized in *S*. Thus, if, as before, *S* is the human body and *G* the normal internal temperature, *G* may be realized in *S* as a consequence of alternate configurations in the causally relevant parts of *S*: for instance, a small flow of blood in the peripheral vessels together with a moderate thyroid secretion and a normal respiration rate may produce the same internal temperature as a greater blood flow combined with a smaller thyroid secretion and a higher respiration rate. Indeed, the possibility of more than one *G* state for *S* is of special interest for the present discussion, since the assumption that *S* is a directively organized system will be seen to require it. But in any event, since *S* is assumed to be a deterministic system with respect to *G*, the *G* state that is realized at one time is uniquely determined by the state of *S* at any previous time – even when the class K_G of the possible *G* states of *S* contains more than one member.

(f) Assuming that *S* is in a *G* state ($A_0B_0C_0$) at time t_0, suppose a change occurs somewhere in *S* so that in consequence the coordinate A_x acquires a different value at the later time t_1, where the times under consideration are in the period *T*. What value it actually does acquire depends on, among other things, the particular structure of S and on what has taken place in it. It may turn out, for example, that A has been made to vary in such a manner that whether or not this change is accompanied by other changes in the remaining state coordinates, S is no longer in a G state. However, let us assume the contrary, and suppose that if A_0 is changed into some other value of the state coordinate A_x, provided that this value belongs to a restricted class of values K_{AG}, B_0 and C_0 will suffer compensating changes which fall into the classes K_{BG} and K_{CG} respectively, so that the value of the matrix $(A_xB_yC_z)$ at time t_1 is a member of the class of G states K_G of S. Thus, we are assuming, for example, that if A_0 becomes A_1 at time t_1, B_0 changes into B_1 and C_0 into C_1, so that $(A_1B_1C_1)$ is a G state; if A_0 becomes A_2, B_0 changes into B_2 while C_0 remains unaltered, so that $(A_2B_2C_0)$ is a G state; and so on. The classes K_{AG}, K_{BG} and K_{CG} are the limits of variation for the state coordinates that are compatible with the conditions generating a G state. Each of these classes is a subclass of the class of permissible values of the corresponding state coordinate – for example, K_{AG} is a subclass of K_A; but in general, the former is a proper subclass of the latter, so that the class of values of the matrix $(A_xB_yC_z)$ which are G states of S does not necessarily coincide with the class K_S of the permissible values of the matrix.

(g) Let us now withdraw the assumption made at the outset that the external environment E is to be ignored. But in withdrawing it, the analysis of functional systems is not essentially altered. For suppose that some factor in E is causally relevant to the occurrence of G in S, and that it is represented by the state coordinate D_w. Then the causally effective states with respect to G of the inclusive system S' (which includes both S and E) will be certain values of the matrix $(A_xB_yC_zD_w)$, and the discussion will proceed as before.

It is worth noting, however, that changes in S usually bring about only relatively insignificant changes in the environmental factors. Accordingly, while some variations in these latter which affect a coordinate for G in S may call forth compensating changes in other coordinates so as to preserve S in a G state, S cannot make such compensating changes for all environmental variations. It is therefore intelligible to talk of the 'degree of plasticity' or the 'degree of adaptability' of an S (say an organism) in relation to its environment, but not conversely.

The discussion has made plain that when a system S is directively organized with respect to some G, the continued existence of G is independent, up to a point, of variations in any of the causally relevant parts of S or factors in E. For despite the fact that G depends on these parts and factors, an alteration in any one of them which might otherwise be fatal to the existence of G is, by hypothesis, compensated by changes in other causally relevant items in S so as to continue S in a G state. If K_{XG} is the set of variations in some causally relevant factor X that can be compensated by changes in S which preserve S in a G state, then the more extensive is K_{XG} the more is the continuance of S in the G state independent of variations in this factor.

(h) We have not assumed anywhere thus far that the state coordinates are numerical variables. However, if this supposition is made, together with the usual assumptions about continuity and the like, we can express the substance of our discussion in the notation of mathematical analysis.[2]

Let S be a system, G a trait of the system, and $x_1, x_2, \ldots, x_n$ the coordinates of state for G. These coordinates are continuous functions of the time, and their values at a particular time will be indicated by superscripts.

(i) We have already shown how to express the assumption that S is a deterministic system (i.e. that the state of S at one time is uniquely determined by its state at some previous time), namely:

$$x_1^t = f_1(x_1^{t_0}, \ldots, x_n^{t_0}, t - t_0)$$
$$\cdots\cdots\cdots$$
$$x_n^t = f_n(x_1^{t_0}, \ldots, x_n^{t_0}, t - t_0)$$

where the f_i are single-valued functions of their arguments, and their derivatives with respect to the time are single-valued functions of the state coordinates but of no other functions of the time. But we shall have no occasion to make use of the second part of this proviso.

(ii) The assumption that because of the special character of S and

2. The mathematical formulation which follows is reproduced almost verbatim from Sommerhoff (1950).

its parts certain restrictions are imposed on the values of the state coordinates, can be expressed by the n inequalities:

$$a_i \leqslant x_i \leqslant b_i \ (i = 1, 2, \ldots, n)$$

or alternatively by

$$x_i \epsilon \Delta x_i \ (i = 1, \ldots, n), \text{ where } \Delta x_i \text{ is some interval.}$$

(iii) The assumption that S is in a G state at a given time t during some period T can be expressed by requiring that state coordinates satisfy a set of equations or conditions. Thus, we stipulate that S will be in a G state at time t during some period T, if and only if

$$g_1 (x_1^t, \ldots, x_n^t) = 0.$$
$$\cdot \quad \cdot \quad \cdot \quad \cdot \quad \cdot \quad \cdot \quad \cdot$$
$$g_r (x_1^t, \ldots, x_n^t) = 0.$$

(iv) The n-tuples $(x_1^t, \ldots, x_n^t)$ which satisfy these equations fall into a certain class K_G; and the values of the individual state coordinates which are members of these n-tuples will therefore also fall into restricted classes. This latter restriction can be expressed as follows:

$$a_i \leqslant a_i^G \leqslant x_i^t \leqslant b_i^G \leqslant b_i \ (i = 1, \ldots, n)$$

or alternatively as

$$x_i^t \epsilon \Delta^G x_i \ (i = 1, \ldots, n), \text{ where } \Delta^G x_i \leqslant \Delta x_i$$

(v) The assumption that if S is in a G state at a given time within a specified period T, and the value of one of the state coordinates is altered, then provided that the change falls into a certain interval the G state is preserved, can be expressed as follows:

Let x_k be the state coordinate in question, $x_k^{t_0}$ a value that satisfies the equations g_j for a G state at some initial time t_0, and $\Delta^G x_k$ a G-preserving interval of change. Then for each $j = 1, 2, \ldots, r$, provided that x_k^t falls into $\Delta^G x_k$,

$$\frac{\partial g_j}{\partial x_k^{t_0}} = \frac{\partial g_j}{\partial x_1^t}\frac{\partial x_i^t}{\partial x_k^{t_0}} + \frac{\partial g_j}{\partial x_2^t}\frac{\partial x_2^t}{\partial x_k^{t_0}} + \ldots + \frac{\partial g_j}{\partial x_n^t}\frac{\partial x_n^t}{\partial x_k^{t_0}} = 0$$

(vi) Finally, the assumption that when x_k varies within the interval $\Delta^G x_k$, there are compensating variations in one or more of the other state coordinates so as to preserve S in the G state, can be rendered as follows:

For at least one j ($j = 1, \ldots, r$) there are s nonvanishing summands in the equations cited in (v), such that $2 \leqslant s \leqslant n$, that is, the summands in these equations do not all vanish, so that in one or more of them (say the jth one) there are at least two summands such that

$$\frac{\partial g_j}{\partial x_i^t} \frac{\partial x_i^t}{\partial x_k^{t_0}} \neq 0.$$

II

With this schematic formulation of the distinctive traits of functional systems before us, we now examine Merton's codification of the 'concepts and problems which have been forced upon our attention by critical scrutiny of current research and theory in functional analysis'. We shall first reproduce in turn each item in his paradigm; identify as far as it is possible to do so the distinctions each contains with elements in the formal analysis of the preceding section; indicate whenever necessary possible ambiguities in his discussion; and show at what points in his paradigm the special subject matter of sociology raises problems and requires distinctions for which there appear to be no counterparts in our generalized formulation of functional systems.

1. The item(s) to which functions are imputed

The entire range of sociological data can be, and much of it has been, subjected to functional analysis. The basic requirement is that the object of analysis represent a *standardized* (i.e. patterned and repetitive) item, such as social roles, institutional patterns, social pressures, cultural pattern, culturally patterned emotions, social norms, group organization, social structure, devices for social control, etc.

Basic query: what must enter into the protocol of observation of the given item if it is to be amenable to systematic functional analysis? (Merton, 1949).

(a) The 'data' and 'items' here mentioned appear to have a status in inquiry analogous to the parts and processes of organisms whose functions are being investigated in biology. If the inquiry is successfully completed, and its findings are formulated, some of these items could presumably be represented by *state coordinates* for some trait of the system.

(b) However, Merton's attention appears to be directed primarily to the preliminary stage of a functional analysis, rather than to the completed outcome of such an inquiry – to the stage at which crude and tentative discriminations are being explored, and gross relations of dependence between the discriminated items are being established. There is usually quite a distance between this exploratory state and the formulation of a satisfactory list of state coordinates for some trait of system. Indeed, a proposed list of state coordinates does not become definitive until an adequate theory (or system of general laws) has been established to account for a given set of traits of the subject matter. It is well known that in the development of a science it is often necessary to add to or subtract from an initial list of state coordinates. For ideally the state coordinates must describe completely the state of a system that is causally relevant to the occurrence of a given property. There are no rules for discovering the appropriate set of coordinates; and there is no assurance whatever that they are contained in a catalogue of miscellaneous items discriminated in a subject matter, no matter how exhaustive such a catalogue may seem to be and no matter how carefully the items are observed and collected. It is in fact by no means the case that the most obvious or even the immediately observable parts or features of a system are the features which correspond to an adequate list of state coordinates; and the history of science supplies ample evidence to show that the pertinent coordinates of a system are often related to matters of direct observation only indirectly.

(c) It has already been explained, and merits some emphasis, that the state coordinates for a given trait of a system must be mutually independent variables – in the sense that their respective values at a given time are not derivable from one another. It is not entirely clear from Merton's statement whether the items he mentions are intended as a possible list of coordinates for some *one* state of a system, or whether they are a juxtaposition of several partial lists for *different* states. If the former is intended, the question whether the items cited satisfy the requirement just noted for state coordinates is a factual issue, upon which the present study has nothing to say. There is, however, some prima facie ground for doubt whether, for example, 'social structure' and 'institutional patterns' – if taken as state coordinates for some one state – do meet this requirement.

(d) A partial though formal answer can be given to Merton's Basic Query, on the assumption that it is addressed to the requirements for protocols of observation relating to the values of state coordinates. The notion of state coordinate would have no significant application to empirical inquiry unless at least the following conditions are fulfilled.

(i) A rule must be specified for each state coordinate (or for certain combinations of them) which connects it (or the combinations) with matters of gross observation, however involved and indirect the connection may be.

(ii) Different hypothetical values of a given state coordinate (or of a given combination of state coordinates) must be discriminable in observation – at any rate, with some degree of approximation. For example, if the coordinate is a distance, one must be able to distinguish observationally between a distance of, say, two and two hundred meters, though perhaps not between a tenth and an eleventh of a millimeter.

(iii) The traits represented by different state coordinates (or different combinations of coordinates) must be distinguishable by some form of observation – for example, what is denoted by 'position' must be distinguishable from what is denoted by 'momentum'.

2. *Concepts of subjective dispositions (motives, purposes)*

At some point, functional analysis invariably assumes or explicitly operates with some conception of the motivation of individuals involved in a social system. As the foregoing discussion has shown, these concepts of subjective disposition are often and erroneously merged with the related, but different, concepts of objective consequences of attitude, belief and behavior.

Basic query: in which types of analysis is it sufficient to take observed motivations as *data*, as givens, and in which are they properly considered as *problematical*, as derivable from other data? (Merton, 1949).

(a) It seems reasonable to suppose that reference is here being made to motives and purposes as causally relevant for the occurrence of some phenomena. Accordingly, the reference is to a *special* state coordinate for some state of a system. However, as a state coordinate 'subjective disposition' is on par with any other coordinates (such as those mentioned in the first number of the paradigm); and it is not evident why it should be listed under a special category in what is ostensibly a *general* paradigm of functional analysis.

Whether, in point of fact, 'subjective disposition' is a suitable state coordinate in the study of social systems, is of course not a formal question, and it can be decided only on the basis of the special facts and established laws of the social sciences.

(b) Moreover, not only can the Basic Query receive no answer in general or formal terms; it must also be cleared of some ambiguities to be quite definite.

(i) If 'subjective disposition' is one state coordinate among others in a system, and since the values of state coordinates at a given time are (by definition) independent of one another, the specific value of this coordinate at a given time will not be 'derivable' from the corresponding values of the other coordinates at that time. In this sense of the Query, observed motivations must be taken 'as data, as givens'.

(ii) On the other hand, still on the assumption that 'subjective disposition' is a state coordinate, the specific character of subjective dispositions at one time must be derivable from the values of the state coordinates at some previous time – provided, of course, that suitable laws for the system have already been established. In this sense of the Query, observed motivations can always be 'properly considered as problematical'.

(iii) However, it may be that 'subjective disposition' is not a suitable variable to count as a state coordinate in a given system, perhaps because it does not enter in that capacity into known laws or theories. Two cases can then be distinguished. (α) Although 'subjective disposition' is not a state coordinate, it may be related by some known laws to variables that are state coordinates. In this case, observed motivations will be 'derivable from other data'. (β) It may be that it is related by no known laws to other variables. In this case, motives and purposes will have to be taken 'as data, as givens'.

(iv) Finally, there is the possibility that two alternate but equivalent analyses (or theories) are available for a given system, in one of which 'subjective disposition' is a state variable but in the other not. Accordingly, when considered within the framework of one mode of analyses, subjective dispositions will be 'givens' or 'derivable', depending on whether one is raising the question mentioned under (i) or under (ii); but when considered within the framework of the second mode of analysis, they could always be properly regarded as derivable from other data – that is, from the values of the state coordinates of the second theory.

3. Concepts of objective consequences (functions, dysfunctions)

We have observed two prevailing types of confusion enveloping the several current conceptions of 'function':

(1) the tendency to confine sociological observations to the *positive* contributions of a sociological item to the social or cultural system in which it is implicated; and

(2) the tendency to confuse the subjective category of *motive* with the objective category of *function.*

Appropriate conceptual distinctions are required to eliminate these confusions.

The first problem calls for a concept of *multiple consequences* and *a net balance of an aggregate of consequences.*

Functions are those observed consequences which make for the adaptation or adjustment of a given system; and *dysfunctions*, those observed consequences which lessen the adaptation or adjustment of the system. There is also the empirical possibility of *nonfunctional* consequences, which are simply irrelevant to the system under consideration.

In any given instances, an item may have both functional and dysfunctional consequences, giving rise to the difficult and important problem of evolving canons for assessing the net balance of the aggregate of consequences. (This is, of course, most important in the use of functional analysis for guiding the formation and enactment of policy.)

The second problem (of confusion between motives and functions) requires us to introduce a conceptual distinction between the cases in which the subjective aim-in-view coincides with the objective consequences, and the cases in which they diverge.

Manifest functions are those objective consequences contributing to the adjustment or adaptation of the system which are intended and recognized by participants in the system;

Latent functions, correlatively, being those which are neither intended nor recognized.

Basic query: what are the effects of seeking to transform a previously latent function into a manifest function (involving the problem of the role of knowledge in human behavior and the problems of 'manipulation' of human behavior)? (Merton, 1949)

(a) Despite the general clarity of Merton's observations in connection with the first problem, several distinguishable things seem to be covered by them.
(i) By the 'function' of an item (or set of items) in S one may understand simply some trait G which that item succeeds in maintaining in S. The item can then be represented as a state coordinate for G, and its function is the preservation of G in S. In this sense of the word, a function of an item is a role it plays in S, so that 'function' is being employed in the first of the three meanings of the word that have been distinguished above (p. 299).
(ii) However, the word could be understood to have a more inclusive meaning, and to refer to some or all of the effects (immediate as well as indirect) that are produced by a change in a state variable, provided only that the change falls into some class K_G of variations which preserve S in a G state for some indicated G. Thus, suppose S to have two Gs, G_1 and G_2, where A_1 and B_1 are state coordinates for the first, and A_2 and B_2 are state coordinates for the second. A change in A_1 may bring forth compensating changes in B_1 so as to maintain G_1; but it may also bring about a variation in A_2 that in turn is followed by a G_2-preserving variation in B_2. All these changes in these several coordinates, and not only the maintenance of G_1 and G_2, will then be 'objective consequences' of the initial change in A_1, 'which make for the adaptation or adjustment of a given system', and may be intended by the phrase 'function of A_1'. In this sense of the word, 'function' is being employed in approximately the third meaning previously distinguished.
(b) In terms of the distinctions of our generalized account of functional systems, a dysfunction can be identified as one of the following kinds of change.
(i) If A is a state coordinate for G in S, and A varies so that despite additional changes in other coordinates for G, the variations fall outside the class K_{AG} of G-preserving changes, and thereby take S out of its G state.
(ii) If A varies so as to remain within the class K_{AG} of G-preserving variations, but for some reason another coordinate B changes so

that its new value falls outside the class K_{BG} (and so takes S out of the G state).

(iii) If, as before, S has two Gs, G_1 and G_2, where A_1 and B_1 are coordinates for the first while A_2 and B_2 are coordinates for the second, and if A_1 varies with compensating variations in B_1 so as to preserve G_1, but induces a variation in A_2 which is not compensated by a G_2-preserving variation in B_2 (so that S moves out of the G_2 state). In this third case, the change in A_1 can perhaps be regarded as only a partially dysfunctional one.

(c) Nonfunctional changes can be specified as follows: A system S will in general possess an indefinite number of properties that are not exhausted by the set of state coordinates for a given G (or class of Gs). Suppose S to possess the class of Gs: $G_1, G_2, \ldots, G_n$ and that X is a property of S which does not belong to any set of state coordinates for these Gs nor is a constituent element in any of the Gs. If a change in X induces no functional or dysfunctional variations in any of these coordinates, the change may be said to be nonfunctional with respect to these Gs. An important point to be observed is that just as the claim that a given change is functional or dysfunctional must be understood as being relative to a specified G (or sets of Gs), so the claim that a change is nonfunctional must similarly be construed as being relative to a specified set of Gs. A change which is nonfunctional with respect to G_1 may be functional, or dysfunctional, or nonfunctional relative to G_2.

(d) A similar relational formulation is required for explicating the sense in which 'an item may have both functional and dysfunctional consequences'. Thus, as just noted, a variation in a coordinate may be functional with respect to G_1, but dysfunctional with respect to G_2. Moreover, a system S may undergo development over a period of time (whether as a consequence of a 'natural' growth or of an altered environment); and it is quite possible that though a variation in an item at one time is functional relative to G_1, a similar variation in that item at another time is dysfunctional with respect to G_1. Should such a case arise, it might become a matter of debate whether one is investigating the 'same' system S at different times, or two different systems S_1 and S_2 which happen to stand to one another in some relation of causal continuity.

It is also pertinent to mention in this connection the possibility that some hierarchy has been established for various *G*s which a system may exhibit. Suppose, for example, that a set of four *G*s is arranged in the order: G_1, G_2, (G_3, G_4), where the first takes precedence over the second, the second over the other two, but G_3 and G_4 have the same rank. If a change in *S* is G_1-preserving but not G_2-preserving, it may perhaps count as 'on the whole' functional rather than dysfunctional. If the change is dysfunctional with respect to G_1, it may count as dysfunctional 'on the whole', even if it is functional relative to G_2. However, if the change is nonfunctional with respect to G_1 and G_2, but is functional relative to G_3 and dysfunctional relative to G_4, it would doubtless be regarded as quite arbitrary to characterize the change as being 'on the whole' one kind rather than another.

(e) Merton's second problem concerns a confusion which involves matters specific to the social sciences, and is not covered by the distinction developed in the generalized account of functional systems. Thus, it has already been noted that 'subjective aim-in-view' (or 'subjective disposition' more generally) is at best a special state coordinate which, in a formal analysis of such systems, counts simply as one coordinate among others. Accordingly, unless 'subjective aim-in-view' is explicitly introduced as a special state coordinate, Merton's distinction between manifest and latent functions is vacuous, and all functions fall under the head of 'latent functions'.

However, if 'subjective aim-in-view' is recognized as a special variable, it becomes possible to formulate the question raised in the Basic Query within the framework of such a slightly enlarged formal analysis. For let A_X be this state coordinate, *B* some other coordinate for a given *G* of *S*; and use the notation A_B to represent the intended and recognized consequences of a certain variation in *B*, also X_B to represent the actual consequences of this change in *B*. To say that the intended and recognized consequences of a given change in *B* is the same as the actual consequences of the change, could then be represented as: $A_B = X_B$; and to say that the intended and recognized consequences of the change is different from the actual ones, could be represented as: $A_B \neq X_B$. The question in the Basic Query may then be put as follows. If at time t_0, $A_B \neq X_B$, but at the subsequent time t_1

$A_B = X_B$, what consequences does the change in the two values of A_X have for the system S or some designated part of it?

4. Concepts of the unit subserved by the function

We have observed the difficulties entailed in *confining* analysis to functions fulfilled for 'the society', since items may be functional for some individuals and subgroups and dysfunctional for others. It is necessary, therefore, to consider a *range* of units affected by the given item: individuals in diverse statuses, subgroups, the larger social system and culture systems. (Terminologically, this implies the concepts of psychological function, group function, societal function, cultural function, etc.) (Merton, 1949)

(a) The two important points that are apparently being made in this number of the paradigm can be identified in the abstract statement of functional systems as follows. In a functional analysis, it is essential to specify (i) the system S which is being investigated, as well as (ii) the G of S which is maintained by indicated items in S.

(b) Enough has been said about the second of these points; but the first, despite its obviousness, perhaps deserves a brief elaboration.

(i) A given item will in general be an element in several systems. Suppose that an item belongs to the three systems S_1, S_2, S_3, where the first is a part of the second, the second a part of the third, and G_1, G_2 and G_3 are Gs in them respectively. Assume that the item is causally relevant to G_1 and G_2, but not to G_3, and that it is represented by the coordinate A. Then a change in A will be nonfunctional with respect to G_3; but it may be functional in relation to both G_1 and G_2, dysfunctional in relation to both, or functional relative to one and dysfunctional relative to the other.

(ii) A given item may be an element in a system S_1, which is part of the environment of another system S_2. If this item is causally relevant to G_1 of S_1, and if S_2 exhibits G_2, a variation in this item may be either functional or dysfunctional with respect to G_1, and at the same time functional, dysfunctional or nonfunctional with respect to G_2.

(iii) A system S_1 that exhibits G_1 may contain two subordinate systems S_2 and S_3 as parts, where these parts exhibit G_2 and G_3

respectively. If an item which is an element of S_2 (and hence of S_1) but not of S_3 is causally relevant to both G_1 and G_2, a variation in it may be functional with respect to all three *G*s, or dysfunctional with respect to all of them, or nonfunctional with respect to G_3 and functional with respect to one of the other *G*s but dysfunctional in relation to the third one.

5. Concepts of functional requirements (needs, prerequisites)

Embedded in every functional analysis is some conception, tacit or expressed, of functional requirements of the system under observation. As noted elsewhere, this remains one of the cloudiest and empirically most debatable concepts in functional theory. As utilized by sociologists, the concept of functional requirement tends to be tautological or *ex post facto*; it tends to be confined to conditions of 'survival' of a given system; it tends, as in the work of Malinowski, to include biological as well as social 'needs'.

This involves the difficult problem of establishing types of functional requirements (universal *v.* highly specific); procedures for validating the assumption of these requirements; etc.

Basic query: what is required to establish the validity of such an intervening variable as 'functional requirement' in situations where rigorous experimentation is impracticable?' (Merton, 1949)

(a) The main point that appears to be noted here is the variety of *G*s which a given system *S* may exhibit, and the importance for a clear-headed functional analysis of specifying which one of the several *G*s is under discussion.

It is of course not an easy matter to establish the complete set of state coordinates for a given *G*. But it is also evident that a proposed list will tend to be used tautologically, as Merton observes, unless the *G* for which they are alleged coordinates is carefully indicated. For a variation in *any* item in *S* will have *some* consequences in *S*; and any item could easily be made to count as a state coordinate if the *sole* ground for such a designation were the fact that its variation produces some changes in *S*.

(b) However, the notion of 'functional requirements' suggests something further – a classification of the various *G*s of a system on the basis of some principle, and perhaps the establishment of a hierarchy among them.

(i) It may be supposed that for a given system *S* there are certain comprehensive *G*s corresponding to the 'vital functions' of biological organisms (i.e. respiration, reproduction, etc.) which are 'indispensable' to the 'survival' of *S*. A list of such *G*s is in effect a *definition* (or part of a definition) of what is to be understood by a given *S*, so that if G_1 is on this list, to say that G_1 is essential to the survival of *S* is to utter a tautology. Now although in principle it is quite easy to construct such a defining list of *G*s, and although in some areas of study (e.g. biology) there is general agreement on the membership of such a list, in other domains, as Merton observes, agreement is difficult to obtain and the specification of such 'vital functions' may remain debatable for an indefinite period.

(ii) In any event, the construction of a typology for the *G*s of a system, or the establishment of a hierarchy among them, requires special material assumptions depending on the *S* that is being considered. A generalized account of functional analysis cannot be expected to resolve these problems. However, it is possible within the framework of such an account to make explicit the pattern of relations that are involved in the distinction between 'universal' and 'highly specific' functional requirements, perhaps in the following manner.

Suppose that $G_1, G_2, \ldots, G_n$ are a set of *n* mutually exhaustive and exclusive *G*s for a given *S*, in the sense that at a given time or period one of them is realized in *S* but that the occurrence of one of them in *S* at a given time excludes the occurrence in *S* at that time of any of the others. Suppose further that the occurrence in *S* of some one G_i ($i = 1, 2, \ldots, n$) at any time implies the occurrence in *S* at that time of G^*, but not conversely. Under the conditions supposed, we may say that G^* is a 'universal' functional requirement for *S* during a specified period, while any of the G_is are 'more specific' functional requirements.

(c) The question raised in the Basic Query does not appear to be specific to functional analysis, and can be pertinently raised in any causal inquiry where 'rigorous experimentation' is precluded. For to ask whether a given item *A* contributes to the maintenance of a specified *G* in *S*, or whether *G* is indispensable to *S*, is either to ask whether the occurrence of *G* is dependent on *A* and its interrelations with other items, or whether some other (defining) traits of S are dependent on the preservation of G. And these are questions which, though they may be difficult to answer, arise in all domains of inquiry, and not only in the study of directively organized systems.

6. Concepts of the mechanisms through which functions are fulfilled

Functional analysis in sociology, as in other disciplines like physiology and psychology, calls for a 'concrete and detailed' account of the mechanisms which operate to perform a given function. This refers, not to psychological, but to social, mechanisms (e.g. role-segmentation, insulation of institutional demands, hierarchical ordering of values, social division of labor, ritual and ceremonial enactments, etc.).

Basic query: what is the presently available inventory of social mechanisms corresponding, say, to the large inventory of psychological mechanisms? What are the methodological problems entailed in discerning the operation of these social mechanisms? (Merton, 1949)

(a) This number of the paradigm is prima facie simply a call for an explicit listing of the state coordinates for the various Gs of social systems, and thus seems in part to be but a restatement of a point already noted in the preceding number.

(b) It should be added, however, that any inquiry into functional systems would doubtless not come to an end with the mere discovery of a complete set of state coordinates for a given G, but would also seek to establish the detailed modes of dependence between the states of the system at different times as well as the specific conditions under which the G occurs. If we employ the notation of the mathematical formulation of directively organized systems given earlier, the point of the present number of the paradigm can then be restated as the following threefold requirement for functional analysis: the specification of the state coordinates $x_1, \ldots, x_n$ for a given G; the formulation of the relations of dependence $f_1, \ldots, f_n$, which hold between the coordinates at different times; and the discovery of the conditions $g_1, \ldots, g_r$ under which the G occurs.

(c) Merton's apparent insistence upon social rather than psychological mechanisms is obviously predicated upon the assumption that a distinction can be drawn between them which is sufficiently clear for the purposes at hand. Moreover, although this is far less certain, he seems to adopt the further material assumption that in sociology an adequate list of state coordinates for a given G will contain only coordinates referring to distinctively social items. But these assumptions, whether or not Merton is actually committed to them, involve factual issues that fall outside the scope of this study.

7. *Concepts of functional alternatives (functional equivalents or substitutes)*

As we have seen, once we abandon the gratuitous assumption of the functional indispensability of given social structures, there is immediately required some concept of functional alternatives, equivalents, or substitutes. This focuses attention on the *range of possible variation* in the items which can, in the given instance, subserve a functional requirement. It unfreezes the identity of the existent and the inevitable.

Basic query: since scientific proof of the equivalence of an alleged functional alternative ideally requires rigorous experimentation, and since this is not often practicable in large-scale sociological situations, which practicable procedures of inquiry most nearly approximate the logic of experiment? (Merton, 1949)

(a) The point here noted is easily recognized as central in the analysis of directively organized systems, and as an expression of the basic idea upon which the generalized account of the preceding section is built. Stated in the notation of that formulation, the point is that if the class K_G of possible G states for a specified G of a system contains more than one member (and this condition may be supposed to be contained in the assumption that the system is directively organized with respect to G), G can occur as a consequence of different configurations of elements in S (or its environment) and in relative independence of the variations in any one of the causally relevant elements.

However, though this point is by now quite familiar, a few further ampliative comments may be useful.

(i) A system may 'grow' or 'develop' in time, so that although a given G may be maintained throughout the development, the state coordinates for G and the mode of the mutual dependence of their values at different times may change. This possibility and its obvious consequences can be made formally explicit as follows. Suppose that $X_1, \ldots, X_n$ is a hypothetically complete list of state coordinates in S for a given G. It may happen that during a certain period T_1, or under certain conditions c_1, G is preserved in S, though the item X_n is inoperative (whether because it does not occur in S in that period or because other circumstances cause it to be simply a 'sleeping partner'). But it may turn out that during a subsequent period T_2, or under other conditions c_2, while G is still preserved in S, X_n is no longer inoperative although X_1 now becomes inactive (whether because of its disappearance from S or because of some induced quiescence). The quiescence of X_n in S during T_1 and its activity during T_2 can be taken in the present context as a formal representation of growth or other modes of development in S; and the inverse supposition for X_1 can be used to represent senescence or other modes of decay in S. But however this may be, the possibility that at different times and under different circumstances different items in S are causally operative for maintaining a given G in S, makes further evident that a given G may be preserved through the operation of different instrumentalities.

(ii) A different though analogous possibility is contained in the supposition that a certain G^* in S is 'universal' (in the sense explained above) in relation to the 'more specific' traits $G_1, \ldots, G_n$. Since in this case the occurrence of G^* is contingent upon the realization of some one of the G_is, but is independent of the realization of any particular G_i, the point stressed in this number of the paradigm immediately follows.

(b) In the generalized account of functional systems presented earlier, explicit recognition was given to the fact that a system S has an environment E. Since E is generally the locus of some of the items represented by the state coordinates for a given G, and therefore contributes something to the range of G-preserving variations in the items causally relevant to G, the recognition of E appears to be of considerable importance for the point stressed in this number of the paradigm.

However, for reasons that are not obvious, Merton makes no explicit mention of the environment in which an object of functional analysis in sociology is embedded, though presumably every such object does have an environment. This reminder of one missing item in Merton's codification applies not only to the present number of his paradigm, but to others as well.

8. Concepts of structural context (or structural constraint)

The range of variation in the items which *can* fulfil designated functions within a given instance is not unlimited (and this has been repeatedly noted in our foregoing discussion). The interdependence of the elements of a social structure limit the effective possibilities of change or functional alternatives. The concept of structural constraint corresponds, in the area of social structure, to Goldenweiser's 'principle of limited possibilities' in a broader sphere. Failure to recognize the relevance of interdependence and attendant structural restraints leads to utopian thought in which it is tacitly assumed that certain elements of a social system can be eliminated without affecting the rest of the system. This consideration is recognized by both Marxist social scientists (e.g. Karl Marx) and by non-Marxists (e.g. Malinowski).

Basic query: how narrowly does a given structural context limit the range of variation in the items which can effectively satisfy functional requirements? Do we find, under conditions yet to be determined, an area of indifference, in which any one of a wide range of alternatives may fulfill the function? (Merton, 1949)

(a) Our generalized account of functional systems has distinguished two kinds of constraints upon the state coordinates of such systems, though only one of them appears to be explicitly recognized in this number of the paradigm.

(i) The structure of a given system *S* imposes certain 'boundary conditions' or general constraints upon the items represented by a set of state coordinates, in virtue of which the values of the coordinates *A*, *B*, etc., must all fall into certain ranges of values K_A, K_B, etc., respectively. Alternatively, the possible states of *S* must fall into a certain class K_S. This type of constraint is not mentioned in the present number of the paradigm.

(ii) Since there are certain conditions which the values of the state coordinates *A*, *B*, etc., must satisfy if they are to be the values that determine a *G* state of *S*, these values must fall into certain restricted classes K_{AG}, K_{BG}, etc. Alternatively, the possible *G* states of *S* must all be members of a restricted class K_G. This is apparently the type of constraint to which Merton is calling attention; and the analogue he suggests for Goldenweiser's principle of limited possibilities is an immediate corollary to it.

(iii) If G^* is universal in *S* with respect to the more specific G_1, . . ., G_n, the latter constitute a set of 'indifferent' alternatives, any one of which entails the realization of G^*. Accordingly, though both types of constraints must be recognized for the coordinates for each of the G_is, the existence of these alternatives in a sense can mitigate the practical force of the restraints by permitting a choice between them and yet preserving G^*.

(b) However, it is conceivable that Merton has in mind more involved forms of constraint that are analysable as compounded out of restraints of the second type. Two of the large number of such more complex forms will be mentioned.

(i) Suppose S to be capable of maintaining two distinct Gs, G_1 and G_2, and that A is a state coordinate for both of them. Suppose, moreover, that though a variation in A is functional with respect to G_1 as long as the change falls into the class K_{AG_1}, it is dysfunctional with respect to G_2 unless the change falls into the narrower class $K_{AG_1G_2}$. Accordingly, for the maintenance of both Gs, more severe restrictions are placed upon the possible variation of an item than for the maintenance of one of them.

Moreover, as a consequence of ignorance as to what is the complete set of state coordinates for G_1, a variation in a known coordinate may produce compensating variations in an unknown one so as to preserve G_1 though at the same time giving rise to certain 'side effects' that are dysfunctional with respect to G_2. Here too, more restrictive limits must be supposed to hold for the variation of the first variable if both Gs are to be preserved.

(ii) On the other hand, though a proposed list of coordinates for a given G may be complete, it may contain redundant items (in the sense that their values at a given time are not mutually independent, so that in effect the proposed coordinates fail to satisfy a requirement for state coordinates). Or again, though a proposed list of coordinates is complete without redundancy, it may contain items that are causally irrelevant to a given G. In either case, mistaken ideas can easily arise as to the limits of possible G-preserving changes in S, in some cases the supposed limits being perhaps wider than the actual ones, in other cases narrower.

9. Concepts of dynamics and change

We have noted that functional analysts *tend* to focus on the statics of social structure and to neglect the study of structural change. The concept of dysfunction, which implies the concept of strain, stress and tension on the structural level, provides an analytical approach to the study of dynamics and change. How are observed dysfunctions contained within a given structure, so that they do not produce instability? Does the accumulation of stresses and strains produce pressure for change in such directions as are likely to lead to their reduction?

Basic query: does the prevailing concern among functional analysts with the concept of *social equilibrium* divert attention from the phenomena of *social disequilibrium*? Which available procedures will permit the sociologist most adequately to gauge the accumulation of stresses and strains in a given social system? To what extent does the structural context permit the sociologist to anticipate the most probable directions of social change? (Merton, 1949)

Several distinguishable though related problems are conceivably covered by this number of the paradigm.

(a) Changes in the coordinates for a given G of S which fall outside the limits of the class K_G of G-preserving changes, are dysfunctional with respect to that G. One problem is therefore the investigation of those circumstances, whether their locus is in S or in E, which bring about such changes.

(b) Changes that are dysfunctional relative to a given G may nevertheless be instrumental for the maintenance or the emergence of some other G (which may be a foreseen or an unforeseen consequence of those changes). Attentive scrutiny of S with a view to discovering such suspected Gs is thus suggested by the occurrence of variations that are dysfunctional with respect to a given G.

(c) The mode of dependence of a set of coordinates upon one another (i.e. the f_is in the mathematical formulation of functional systems) may change with time, whether as a consequence of alterations in other items of S or in E; and such a change may give rise to alterations in the conditions under which a given G can occur (i.e. in the G_is of the mathematical formulation). Accordingly, the class K_G of G-preserving variations for a given G may not remain constant. Should K_G shrink to nothing, the given G will no longer be realizable in S; should it become enlarged, G will be capable of being maintained under a more flexible set of circumstances than previously. The discussion therefore suggests the study of possible variations in the conditions under which a given G can occur.

(d) If G_1 and G_2 are related to G^* as more specific functions to a universal one, a change in a coordinate which is dysfunctional with respect to G_1 may nevertheless eventuate in the occurrence of G_2. Accordingly, S will remain stable with respect to G^*, despite an initial dysfunctional change relative to G_1. The problem is thus suggested whether prima facie dysfunctional changes in a system may nevertheless not be entirely compatible with the preservation of some assumed 'vital function' of the system.

(e) A system S may exhibit at different times a series G_1, G_2, . . . of mutually incompatible Gs which succeed each other because of certain 'built in' features of S or because of certain progressive changes in E or both. The double problem then would be to (i) ascertain the order of the succession of the G_is, with the aim of formulating their law of development, and (ii) discover the state coordinates which control the development.

10. Problems of validation of functional analysis

Throughout the paradigm, attention has been called repeatedly to the *specific* points at which assumptions, imputations and observations must be validated. This requires, above all, a rigorous statement of the sociological procedures of analysis which most nearly approximate the *logic* of experimentation. It requires a systematic review of the possibilities and limitations of *comparative* (cross-cultural and cross-group) *analysis.*

Basic query: to what extent is functional analysis limited by the difficulty of locating adequate *samples of social systems* which can be subjected to comparative (quasi-experimental) study? (Merton, 1949)

The questions here raised are factual ones, requiring for their answer familiarity with currently available techniques of sociological study. Nothing said in this paper can throw any light upon them.

11. Problems of the ideological implications of functional analysis

It has been emphasized in a preceding section, that functional analysis has no intrinsic commitment to a given ideological position. This does not gainsay the fact that *particular* hypotheses advanced by functionalists may have an identifiable ideological role. This, then, becomes a specific problem for the sociology of knowledge: to what extent does the social position of the functional sociologist (e.g. *vis-à-vis* a particular 'client' who has authorized a given research) evoke one rather than another formulation of a problem, affect his assumptions and concepts, and limit the range of inferences drawn from his data?

Basic query: how does one detect the ideological tinge of a given functional analysis and to what degree does a particular ideology stem from the basic assumptions adopted by the sociologist? Is the incidence of these assumptions related to the status and research of the sociologist?

(a) These questions again deal with substantive matters, and they are obviously relevant to all sociological inquiry, and not to functional analysis in sociology exclusively.

(b) It is perhaps worth noting at this place that the commitment of a functional analyst in sociology to some 'ideological position' is quite innocuous, if the analyst makes clear what *G* of a system he is investigating, and if he indicates explicitly to the maintenance of what *G* a given item in a system allegedly contributes. Functional analyses in all domains, and not only in sociology, run a similar risk of dogmatic provincialism which characterizes some analyses in sociology, when the relational character of functional statements is ignored, and when it is forgotten that a system may exhibit a variety of *G*s or that a given item may be a member of a variety of systems.

References

ASHBY, W. R. (1952), *Design for a Brain*, Wiley.
CANNON, W. B. (1932), *The Wisdom of the Body*, Norton, New York, and Routledge, London [rev. edn 1963, Norton].
MERTON, R. K. (1949), *Social Theory and Social Structure*, Free Press.
NAGEL, E. (1953), 'Teleological explanation and teleological systems', in S. Ratner, ed., *Vision and Action*, Rutgers University Press.
SOMMERHOFF, G. (1950), *Analytical Biology*, Oxford University Press.
WOODGER, J. H. (1929), *Biological Principles*, Harcourt, Brace [rev edn 1966, Humanities Press].

16 R. L. Ackoff

Systems, Organizations, and Interdisciplinary Research

R. L. Ackoff, 'Systems, organizations, and interdisciplinary research', *General Systems Yearbook*, vol. 5 (1960), Society for General Systems Research, pp. 1–8.

'Speak English!' said the Eaglet, 'I don't know the meaning of half those long words, and, what's more, I don't believe you do either!' And the Eaglet bent down its head to hide a smile: some of the other birds tittered audibly. (*Alice's Adventures in Wonderland* by Lewis Carroll)

Introduction

When the announcement was made of the establishment of a Systems Research Center at Case, a number of my associates in Operations Research asked me how the activity of the Center was to differ from that of the Operations Research Group at Case. My colleagues in the Control (or Systems) Engineering Group at Case were asked similar questions concerning the relationship of their group to the Center. The question could be answered by saying that the Center is designed to facilitate cooperative research and educational activities among the Operations Research Group, the Control Engineering Group, and other systems-oriented activities at Case, particularly the Computing Center.

This answer may satisfy the curiosity of some and discourage probing by others. It is not enough, however, to satisfy or discourage probing by those of us who have some responsibility for the development of this Center. Much more than good will among men is required to make this Center play a significant role in research and education. Part of what is required is a philosophy and a program. A philosophy and program for the Center cannot be expected to spring into existence in a mature state; it must evolve out of proposals, discussion, reformulations, and experience. I should like here to formulate an initial philosophy and program which I hope will lead to constructive discussion, not only at this conference, but afterwards in other cells in which the systems movement is taking shape. There is no doubt in my mind that centers such as we are forming here at Case will develop in profusion in other academic institutions and in industrial and governmental organizations.

I will use my own interdiscipline, Operations Research, as my springboard. But before I take the leap I would like to make some general observations about the systems movement.

First, I believe the systems movement will reach its fruition in an interdiscipline of wider scope and greater significance than has yet been attained. I should like to emphasize that my concern is not with what systems research is, but rather with what we can make of it. I consider operations research as an intermediate step toward this fruition, a step away from traditional science. Correspondingly, I take systems engineering to be an intermediate step toward the same objective, a step away from traditional engineering. I believe systems engineering and operations research are rapidly converging. What more fitting title for the convergence than 'systems research'.

Operations Research is concerned with increasing the effectiveness of operations of organized man–machine systems. A complete understanding of the significance of this too brief characterization requires at least definitions of system, operations, and organization. I shall deal with the first two very lightly,[1] only enough for my immediate purposes. I shall, however, deal with the concept of organization in more detail because I shall use it as the key to the philosophy and program which I hope to develop. It is in the context of organized man–machine systems, I believe, that we find the most comprehensive demands for departure from the existing content and structure of science and technology.

Now, to the task.

Systems and Operations

The term 'system' is used to cover a wide range of phenomena. We speak, for example, of philosophical systems, number

1. For a more detailed discussion of 'systems' and 'operations', see Ackoff (1961).

systems, communication systems, control systems, educational systems, and weapon systems. Some of these are conceptual constructs and others are physical entities. Initially we can define a system broadly and crudely as *any entity, conceptual or physical, which consists of interdependent parts.* Even without further refinement of this definition it is clear that in systems research we are interested only in those systems which can display activity – that is *behavioral* systems.

It is also apparent that systems research is only concerned with behavioral systems which are subject to control by human beings. Consequently, the solar system – although it may be on the verge of becoming so – is not yet a part of the subject matter of systems research. The relevant domain of such research, then, is controllable behavioral systems.

The essential characteristic of a behavioral system is that it consists of parts each of which displays behavior. Whether or not an entity with parts is considered as a system depends on whether or not we are concerned with the behavior of the parts and their interactions.

A behavioral system, then, is a conceptual construct as well as a physical entity since such a system may or may not be treated as a system, depending on the way it is conceptualized by the person treating it. For example, we would not normally think of a man who starts the car as a system because we do not distinguish the parts of the man involved in the component acts. We may, however, consider man as a biological system when studying the metabolic process. A physical entity is considered as a system if the outcome of its behavior is conceptualized as the product of the interactions of its parts. Therefore, many entities may be studied either as elements or as systems; it is a matter of the researcher's choice.

The behavior displayed by a system consists of a set of interdependent acts which constitute an operation. *Operation* is a complex concept which I do not want to deal with in detail here. Loosely put, a set of acts can be said to constitute an operation if each act is necessary for the occurrence of a desired outcome and if these acts are interdependent. The nature of this interdependence can be precisely defined. The relevant outcome and acts involved in an operation can each be defined by a set of properties which can be treated as variables. The acts are interdependent relative to the outcome if the rate of change of any outcome variable affected by change in any variable describing one of the acts depends on (i.e. is a function of) all the other relevant act-variables. Therefore, if the variables can all be represented by continuous quantities, the derivative of an outcome-variable with respect to any act-variable (if it exists) is a function of all other act-variables. In ordinary language, then, an outcome is the product of a set of interdependent acts if it is more than the sum of (or difference between) these acts.

Organization

An organization can be defined as an at least partially self-controlled system which has four essential characteristics:

1. *Some of its components are animals.* Of particular interest to us, however, are those systems in which the animals are human beings. Wires, poles, switchboards, and telephones may constitute a communication system, but they do not constitute an organization. The employees of a telephone company make up the organization that operates the communication system. Men and equipment together constitute a more inclusive (man–machine) system that we can refer to as organized. Since most organizations utilize machines in a significant way in order to achieve their objectives, the discussion here will be directed to organized man–machine systems.

2. *Responsibility for choices from the sets of possible acts in any specific situation is divided among two or more individuals or groups of individuals.* Each subgroup (consisting of one or more choices of action and the set of choices is divided among two or more subgroups. The classes of action and (hence) the subgroups may be individuated by a variety of types of characteristics; for example:

(a) *by function* (e.g. the department of production, marketing, research, finance, and personnel of an industrial organization),

(b) by *geography* (e.g. areas of responsibility of the Army), and

(c) *by time* (e.g. waves of an invading force). The classes of action may, of course, also be defined by combinations of these and other characteristics.

It should be noted that the individuals or groups need not carry out the actions they select; the actions may be performed by machines or other human beings which are programmed or controlled by the individuals so that they act as desired. It should also be noted that the equipment involved and the subgroups may also be considered as systems; that is, as subsystems.

3. *The functionally distinct subgroups are aware of each other's behavior either through communication or observation.* In many laboratory experiments, for example, subjects are given interrelated tasks to perform and are rewarded on the basis of an outcome which is determined by their collective choices. The subjects, however, are not permitted to observe or communicate with each other. In such cases the subjects are unorganized. Allow them to observe each other or communicate and they may become an organization. Put another way, in an organization the human subgroups must be capable of responding to each other either directly or indirectly.

4. *The system has some freedom of choice of both means (courses of action) and ends (desired outcomes).* This implies that at least some parts have alternative courses of action under at least some possible sets of conditions. The simplest type of system, the binary type, has only two possible states: 'off' and 'on' (e.g. a heating system in a home). More complex *adaptive* systems can behave differently under different conditions, but only in one way under any particular set of conditions (e.g. a ship operated by automatic pilot). Still others are free to choose their means to an end but have no choice of this end (e.g. a computer programmed to play chess). Finally, there are those which are free to choose *how* they will act in any situation (means-free) and why (ends-free). To be sure, such systems are usually constrained in their choices by larger systems which contain them (e.g. government restrictions on a company's behavior). Their efficiency is also affected by either the behavior of other systems (e.g. competition in industry) or natural conditions (e.g. weather).

The four essential characteristics of an organization, then, can be briefly identified as its content, structure, communications, and decision-making (choice) procedures.

Design and Operation Organized (Man–Machine Systems)

Now we want to consider the significance of these characteristics to one who wants either to create an effective organized system or to improve the operation of an existing one. He has four basic types of approach to organizational effectiveness and combinations thereof. The basic types of approach correspond to the four essential characteristics of organizations.

Content

The content (men and machines) of an organization can be changed. The study of organizational personnel – their selection, training, and utilization – has come to be the domain of *industrial psychology* (Haire, 1959).

Three fundamentally different approaches to personnel problems have developed within industrial psychology. The first, *personnel* psychology, is primarily concerned with selecting the right man for a specified job. Its principal activity, therefore, is directed toward specifying the relevant characteristics of a job, determining which individual properties are related to its performance, and selecting those individuals who are best equipped for the job. The personnel psychologist, therefore, takes the task-to-be-done as fixed and varies the men.

The personnel psychologist is also interested in modifying man so that he is better capable of performing the task. He attempts such modification through education and training. Here he partially overlaps with the *industrial engineer* who tries to modify the behavior of man more directly. On the basis of time and motion studies the industrial engineer attempts to find those movements which optimize the individual's operations. Industrial engineers, therefore, are preoccupied with manual operations whereas the personnel psychologist tends to concentrate on communication and decision making.

The second psychological approach is that of the human engineer. The human engineer tries to modify the job-to-be-done so that it can be done better by the people available to do it. Here the men are taken to be fixed and the task is taken to be variable. Hence, human engineers, like industrial engineers, are concerned with the acts to be performed, but they try to modify them through the design of the equipment involved in these tasks. It is only natural, therefore, that there has been an increasing convergence of these two approaches.

A third psychological approach takes both the man and the job to be fixed, but the psychological and social environment to be variable. This type of approach yields studies of motivation, incentive systems, interpersonal relationships, group identification or alienation, and the like, and the effect of such variables on human productivity, job satisfaction, and morale. These studies are essentially social-psychological in nature and are epitomized by the early work of Mayo and Roethlisberger and Dickson. Studies of the social environment frequently consider the effect of the non-content aspects of organization (structure, communication, and control) on human performance. For example, the effect of various types of communication networks on the performance of an individual in the network has been extensively explored. Clearly, such studies are related to ones directed at structure, communication, and control; but the emphasis of most of them is on the *individual's* performance rather than on the performance of the organization as a whole.

The other part of the content of man–machine systems is equipment. We have already observed that human engineers are concerned with modifying equipment so that it can be better operated by available personnel. They seldom, however, completely design this equipment. Normally they collaborate with representatives of the traditional branches of engineering in design activity so that the latter can take the capabilities of the operators into account more effectively. Human engineers, therefore, do not replace, they supplement the traditional engineer in his design function.

The individual piece of equipment can frequently be studied as a system. Engineers have increasingly tended to so regard the machine and the weapon. In equipment incorporating automatic controls the systems approach is almost inescapable. In addition, engineers have become increasingly concerned with the interactions of equipment in machine and weapon complexes and so they have become concerned with larger and larger equipment systems. Out of this concern the interdiscipline of *systems engineering* has emerged. The engineer, of course, can no more ignore the human operator than the personnel psychologist can ignore the machine to be operated. The variables which they manipulate, however, remain distinct.

Structure

The second major approach to organizational effectiveness is through its structure, that is, to the way that the necessary physical and mental labor is divided. Although political scientists, economists, and sociologists have concerned themselves with organizational structure, there is as yet no organized body of theory or doctrine of practice on which a unified disciplinary or interdisciplinary applied-research activity can be based. As a consequence most studies of organizational structure, such as those leading to reorganization of a system, are generally done by managers or management consultants whose approach involves more art and common sense than science.

Within the last few decades there has been increasing experimental study of organizational structure. More recently there has begun to appear a body of mathematical theory of organizational structure. Haire has pointed out, however, that as yet,

> We do not have much in the way of systematic behavioral data collected for the purpose of testing hypotheses or quantifying variables used in models. For example, we have models dealing with the cost of decentralized decision-making in abstract terms, but we know nothing about the information and decision load that can be supported, or how individuals vary along this dimension . . . We know little about the effect of various communication structures and practices on alternative forms of organizations and their cohesiveness . . . We should be just on the brink of a period of exciting systematic data collection. (Haire, 1959, p. 72)

We may have reached that brink in a provocative new development: operational gaming (Thomas and Deemer, 1957; Guetzkow, 1957). In operational gaming organized groups are given problems analogous to real ones, usually with a collapse of the time dimension, and are observed under controlled conditions. We appear to be developing a way of experimenting quantitatively with at least small organizations under conditions which appear to be relevant to actual operations (Clark and Ackoff, 1959). Difficult problems remain concerning inferences from the game to the real situation, but there is little doubt that within the next few years significant reduction of these difficulties will occur.

Communication

The effectiveness of an organization depends in part on its having 'the right information at the right place at the right time'. The study of organizational communication is in much the same stage of development as the study of organizational structure. It has no organized body of theory, but it has been developing a doctrine of practice. *Systems and procedures analysts*, stimulated by the numerous installations of automatic data processing systems, have been perfecting techniques of qualitative analysis of information and its flow. It may seem peculiar that this work is predominantly quantitative in light of the highly developed mathematical theory of communication (based to a large extent on the work of Claude Shannon) and its pervasive application to the design of physical communication and information-processing systems. (See Shannon and Weaver, 1949.)

This theory, however, concerns itself exclusively with the physical aspects of communication and has no relevance to problems involving the meaning of the communication. In Shannon's theory, for example, the measure of information contained in a message is a function of the number of distinct physical messages that could have been sent and the probability associated with the selection of each. The measure makes no reference to the content or significance of the message.

The same thing has been said very well by Haire in his discussion of an article by Rapoport:

He [Rapoport] points out that in dealing with communication among linked individuals we have tended to use the information theory on 'bits' developed for the communication engineering. Such a formulation is useful for determining channel capacity . . . but it is not maximally useful for studying decision-making in groups. Here one needs a model of the cognitive aspects of communication theory – a way to indicate the potential of bits for reducing uncertainty about a real state of affairs. Such an approach contrasts with the definition of information in terms of the probabilities of selecting a certain class of messages from a source with given statistical characteristics (Haire, 1959, p. 7).

There is a growing body of experimental work on the effect of different types of communication networks on organizational (rather than individual) performance, particularly on small groups. (Such experimentation has been stimulated to a large extent by the pioneering work of Alex Bavelas, 1950.) In addition, the body of special theories is rapidly expanding so that we may well be on the verge of a major break-through in this area. (This work is very effectively summarized in the recent work of Colin Cherry, 1957.)

Beginnings toward the construction of a behavioral theory of communication have been made here at Case (Ackoff, 1958). This theory has two essential characteristics. First, it does not equate the transmission of information with communication but recognizes three types of message content: information, instruction, and motivation. Information is defined and measured in terms of the effect on the receiver's possibilities and probabilities of choice. Instruction is defined and measured in terms of the effect on the efficiency of the receiver's action, and motivation in terms of the effect of the message on the values which the receiver places on possible outcomes of his choices. A single message may combine all three types of content.

The second essential aspect of this theory is that it provides separate measures of the amount and value of information, instruction, and motivation contained in a message. It therefore distinguishes between information and misinformation, effective and ineffective instruction and motivation.

A theory with these characteristics, whether the one developed at Case or another, increases the possibility of useful quantitative treatment of organizational communication problems.

Decision making

The last type of approach to organizational problems involves its decision-making procedures. An organization with good personnel and equipment, and an effective structure and communication system may still be inefficient because it does not make effective use of its resources. That is, the operations of the organization may not be efficiently controlled. Control is a matter of setting objectives and directing the organization toward them. It is obtained by efficient decision making by those who manage the operation.

Study of the effective utilization of economic resources in industrial and public organizations is a well-established domain of interest to that splinter group in economics which concerns itself with *micro-economics* and *econometrics*. In the last decade it has produced a rapidly expanding body of theory and research techniques. Concurrent with this development there has been another which deals with a broader class of resources than do the economists alone and, consequently, with a wider variety of organizational decision-problems. This broader interdisciplinary approach to organizational control has come to be known as *operations research*.

The essential characteristics of this interdisciplinary activity lie in its methodology. Out of an analysis of the desired outcomes, objectives of the organization, it develops a measure of performance (P) of the system. It then seeks to model the organization's behavior in the form of an equation in which the measure of performance is equated to some function of those aspects of the system which are subject to management's control (C_i) and which affect the desired outcome, and to those controlled aspects of the system (U_j) which also affect the outcome.

Thus the model takes the form:

$$P = f(C_i, U_j)$$

From the model, values of the control variables are found which maximize (or minimize) the measure of the systems performance:

$$C_i = g(U_j)$$

The solution, therefore, consists of a set of rules, one for each control variable, which establishes the value at which that variable should be set for any possible set of values of the uncontrolled variables. In order to employ these rules it is necessary to set up procedures for determining or forecasting the values of the uncontrolled variables.

It will be recognized that this procedure is one by which equipment systems should be ideally designed. In design one should also develop a consolidated measure of system performance, and identify the variables which the designer can control as well as those uncontrolled aspects of the system or its environment which will affect its performance. Unfortunately, in many cases such a model of a desired equipment system cannot be constructed because of our ignorance. For example, I have not yet seen a good single consolidated measure proposed for the performance of an aircraft. Nor is there sufficient knowledge to relate any of the less perfect available measures of performance to the large number of design variables of such craft. As a consequence, design is currently accomplished by a combination of scientific analysis, intuition, and esthetic considerations. It should be recognized, however, that current design procedures are only an evolutionary stage which will be replaced as rapidly as possible by effective modelling and the extraction of solutions from the resulting models.

Integrated Research into Organized Man–Machine Systems

As we have seen, there is a large group of disciplines and interdisciplines dedicated to studying various aspects of organized man–machine systems. The fact that the subject is so dissected leads to several residual problems. Suppose that an organizational problem is completely solvable by one of the disciplines we have considered. How is the manager who controls the system to know which one? Or, for that matter, how is a practitioner of any one discipline to know in a particular case if another discipline is better equipped to handle the problem than is his? It would be rare indeed if a representative of any one of these disciplines did not feel that his approach to a particular organizational problem would be very fruitful, if not the most fruitful. The danger that results can perhaps best be illustrated by a report that may be apocryphal but which makes the point very well.

The manager of a large office building received an increasing number of complaints about the elevator service in the building. He engaged a group of engineers to study the situation and to make recommendations for improvements if they were necessary. The engineers found that the tenants were indeed receiving poor service and considered three possible ways of decreasing the average waiting time. They considered adding elevators, replacing the existing ones by faster ones, and assigning elevators to serve specific floors. The latter turned out to be inadequate and the first two were prohibitively expensive to the manager. He called together his staff to consider the report by the engineers. Among those present was his personnel director, a psychologist. This young man was struck by the fact that people became impatient with a wait which seemed so short to him. On reflection he became convinced that their annoyance was due to the fact that they had to stand inactive in a crowded lobby for this period. This suggested a solution to him which he offered to the manager, and because it was so inexpensive the manager decided to try it. Complaints stopped immediately. The psychologist had suggested installing large mirrors on the walls of the lobbies where people waited for the elevators.

Those who have worked with systems can recall many such incidents, that is, many except those in which they played a role similar to that of the engineers in the one just recounted. There is undoubtedly a considerable waste of research effort and a considerable failure to obtain successful solutions to system problems just because the wrong discipline was involved. How can this be avoided? We will return to this question in a moment, but now let's look at the second residual problem.

In most problems involving organized man–machine systems each of the disciplines we have mentioned might make a significant improvement in the operations. But as systems analysts know, few of the problems that arise can adequately be handled within any one discipline. Such systems are not fundamentally mechanical, chemical, biological, psychological, social, economic, political, or ethical. These are merely different ways of looking at such systems. Complete understanding of such systems requires an integration of these perspectives. By integration I do not mean a synthesis of results obtained by independently conducted undisciplinary studies, but rather results obtained from studies in the process of which disciplinary perspectives have been synthesized. The integration must come during, not after, the performance of the research.

We must stop acting as though nature were organized into disciplines in the same way that universities are. The division of labor along disciplinary lines is no longer an efficient one. In fact, it has become so inefficient that even some academic institutions have begun to acknowledge the fact. What can be done about it?

If the various disciplines involved in studying systems are brought together organizationally, this would help solve the first type of problem because it would then be possible to have each discipline examine each problem that arises. Presumably, by discussion the interdisciplinary group could determine which of the disciplines is best suited to handle a particular problem if it could be handled exclusively by one discipline.

This type of proximity between the disciplines is not enough, however, to effect a truly interdisciplinary approach to systems. The various disciplines must be able to work together effectively on the problem, not merely before and after the problem is studied. To accomplish this some specific steps should be taken.

First, it will be necessary to construct mathematical models of systems in which content, structure, communication, and decision variables all appear. For example, several cost variables are usually included in a typical operations research model. These are either taken as uncontrollable or as controllable only by manipulating such other variables as quantity purchased or produced, time of purchase or production, number and type of facilities, and so on. These costs, however, are always dependent on human performance, but the relevant variables dealing with personnel, structure, and communication seldom appear in such models. To a large extent this is due to the lack of operational definitions of many of these variables, and, consequently, to the absence of suitable measures in terms of which they can be characterized.

In order to be able to construct truly interdisciplinary models of systems, then, it will be necessary to relate conceptually the variables dealt with in each of the disciplines which should be involved in systems research. This is a formidable task, but a beginning has been made. At the Institute of Experimental Method (which operated at the University of Pennsylvania in 1946 and 1947) a monograph was produced which attempted an interdisciplinary conceptual system (Churchman and Ackoff, 1947). More recently Rudner and Wolfson (1958) have been extending this work, particularly as it applies to organizations.

The second requirement is for the self-conscious development of a sound methodology for systems research. This can be accomplished by turning systems research in on itself since systems research is itself an operation performed by man–machine systems.

The methods and techniques of the traditional sciences and technologies are not good enough for the job which must be done. Let me illustrate this point briefly by reference to only one of many methodological problems that might be discussed.

The performance objectives of most systems can be stated in terms of a number of variables. For example, in a truck we seek such characteristics as speed, rapid rate of acceleration, long range, large pay-load, low cost of operation, and so on. We cannot really optimize the design of an aircraft unless we can in some way amalgamate these performance considerations into a single measure of performance. In a production system, for example, we may have to amalgamate a measure of cost, a measure of the length of time required to fill orders, and a measure of the frequency and duration of shortages. In order to accomplish such an amalgamation of measures we must be able to transform all the scales involved into some common (standard) scale. We have much to learn about how to find the appropriate transformations or 'trade-off' functions.

The criterion of 'best performance' can be shown to depend further on our ability to find the relative value (or utility) of increments along the scales used to measure performance. For example, to a man who is destitute, the value of twenty dollars is clearly not twice that of ten dollars. If it were he would prefer a 51 per cent chance of getting twenty dollars to the certainty of getting ten dollars. We have experimental evidence that this is not the case. It is important, therefore, to increase our ability to measure the value of increments of performance along whatever scale(s) they are measured.

The third requirement for effective systems research is effective education and organization of representatives from practically all the scientific and technological disciplines. Systems research will not be the only beneficiary of such education and organization. The contributing disciplines will be significantly benefited. It is not accidental that so much of the important work currently going on in many disciplines is being done by persons trained in other disciplines. For example, the most important work being done in the behavioral sciences, in my opinion, is that being reported in the two new journals, *Behavioral Science* and *Conflict Resolution.* Most of the contributors to these journals were not trained in the behavioral sciences. Major work in learning theory, for example, is being done by Merrill M. Flood (1954), a mathematician at the University of Michigan, and Frederick Mosteller (Bush, Mosteller and Thompson, 1954), a statistician at Harvard. On the other hand, measurement theory, which has been thought of as the domain of physics since the work of Norman Campbell (revised 1957) has been significantly extended by psychologists such as S. S. Stevens (1951) and Clyde Coombs (Coombs, Raiffa and Thrall, 1954) and philosophers of science such as C. West Churchman (Churchman and Ratoosh, 1959).

The implications of these observations to the educational process are important.

No single individual can be educated so as to be expert in all the disciplinary approaches to systems. It is difficult enough to make him expert in one. We can, however, educate him to an awareness of what others know and can do in systems work and motivate him to desire to work collaboratively with them. Scientific snobbery must go. Systems research cannot thrive where it prevails.

It is my feeling that the two most important steps that can be taken to break down barriers to effective interdisciplinary collaboration are:

(a) to elevate those trained in each discipline to a uniformly high level of competence in mathematics and statistics, and

(b) to educate all students in science and technology to a thorough understanding of scientific method in its most general sense.

Mathematics is the language of science and like all languages it molds the concepts and thinking processes of those who are familiar with it. In my opinion, the behavioral sciences are less mathematically oriented than are the physical sciences not so much because of the difference between the types of phenomena they study as because of the difference of language in which their practitioners think about these types of phenomena. On the other hand, existing mathematics is not adequate to provide a complete basis for quantification in the behavioral sciences because it was developed as a handmaiden of the physical sciences. The greatest challenges to mathematics, it seems to me, are increasingly to be found in the behavioral rather than the physical sciences.

Through an exposure to the accomplishments and problems of scientific method, the student can best come to understand the underlying unity of science and hence of its disciplines. Only by a thorough analysis of research procedures in each of the sciences can one come to an appreciation of the interdependence of the sciences. For example, in this way a student can come to realize that progress in physical science involves (among other things) continuous reduction of observer errors and that perceptual psychology and human engineering have a great deal to contribute to such error reduction. He can also come to understand that the social environment of a physical laboratory affects the reliability of measurements of even simple physical quantities. Through the study of scientific method, then, he can begin to see the scientific crusade for the reduction of error as one that is necessarily interdisciplinary in character.

For the systems researcher methodological self-consciousness has an added importance because, as already observed, research itself is frequently an operation performed by an organized system. As such it is susceptible to the same kind of analysis as are other systems. The possibility of so studying research holds great promise for future increases in research effectiveness in all areas of science and technology.

In summary, then, if systems research is to develop the capacity to conduct effective research on complex as well as simple types of systems we must do the following:

1. Develop a conceptual system which relates the concepts applied to systems by various disciplines and reduce them to quantities which are measurable along compatible scales.

2. Develop methodology better adapted to unique aspects of systems research.

3. Design and put in operation an educational program which produces the kind of researcher who can conduct systems research in an interdisciplinary context.

The era of systems research – and I think this is an era – can become one in which not only science is effectively reorganized, but one in which the educational process is similarly reorganized.

The exciting and challenging character of systems research, then, is not to be found so much in what it is, but in what can be made of it and the research and educational institutions that house it.

References

ACKOFF, R. L. (1958), 'Toward a behavioral theory of communication', *Management Science*, vol. 4, pp. 218–34.

ACKOFF, R. L. (1961), 'The meaning, scope, and methods of operations research', *Progress in Operations Research*, vol. 1, Wiley.

BAVELAS, A., (1950), 'Communication patterns in task-oriented groups', *Journal of the Acoustical Society of America*, vol. 22, pp. 725–30.

BUSH, R. R., MOSTELLER, F., and THOMPSON, G. L. (1954), 'A formal structure for multiple-choice situations', in R. M. Thrall, C. H. Coombs and R. L. Davis (eds.), *Decision Processes*, Wiley.

CAMPBELL, N. R. (1957), *Foundations of Science*, Dover Publications. (Formerly titled *Physics the Elements.*)

CHERRY, C. (1957), *On Human Communication*, M.I.T. Press.

CLARK, D. F., and ACKOFF, R. L. (1959), 'A report on some organizational experiments', *Operations Research*, vol. 7, pp. 279–93.

CHURCHMAN, C. W., and ACKOFF, R. L. (1947), *Psychologistics*, University of Pennsylvania Research Fund (mimeographed).

CHURCHMAN, C. W., and RATOOSH, P. (1959), *Measurement: Definitions and Theories*, Wiley.

COOMBS, C. H., RAIFFA, H., and THRALL, R. M. (1954), in R. M. Thrall, C. H. Coombs and R. L. Davis (eds.), *Decision Processes*, Wiley.

FLOOD, M. M. (1954), 'Game-learning theory and some decision-making experiments', in R. M. Thrall, C. H. Coombs and R. L. Davis (eds.), *Decision Processes*, Wiley.

GUETZKOW, H. (1957), 'The development of organizations in a laboratory', *Management Science*, vol. 3, pp. 380–402.

HAIRE, M. (1959), 'Psychology and the study of business: joint behavioral sciences', in R. A. Dahl, M. Haire and P. F. Lazarsfeld (eds.), *Social Science Research on Business: Product and Potential*, Columbia University Press.

RUDNER, R. S., and WOLFSON, R. J. (1958), *Notes on a Constructional Framework for a Theory of Organizational Decision Making*, Working Paper no. 3, Management Science Nucleus, Institute of Industrial Relations, University of California.

SHANNON, C. E., and WEAVER, W. (1949), *The Mathematical Theory of Communication*, University of Illinois Press.

STEVENS, S. S. (1951), 'Mathematics, measurements, and psychophysics', *Handbook of Experimental Psychology*, Wiley.

THOMAS, C. J., and DEEMER, W. L. (1957), 'The role of operational gaming in operations research', *Operations Research*, vol. 5, pp. 1–27.

Part Five **Systems Management**

The thinking reflected in the earlier parts has already begun to have an impact upon the common notions of management policy making. These selections are primarily intended to give a glimpse of these developments. However, the Hirschman and Lindblom article (Reading 17) has a special interest of its own. While the principle guiding management of separate but correlated systems seems to be that of 'joint optimization', their evidence seems to suggest that management of the relations between, and the performance of the parts of each system needs to be guided by the principle of 'the leading part'. That is, that the ends of the system are best met by optimizing conditions for the part which is for the time the leading part, even at the expense of other parts. Ways' article (Reading 18) is frankly speculative, not theoretical.

17 A. O. Hirschman and C. E. Lindblom

Economic Development, Research and Development, Policy Making: Some Converging Views

A. O. Hirschman and C. E. Lindblom, 'Economic development, research and development, policy making: some converging views', *Behavioral Science*, vol. 7 (1962), pp. 211–22.

When, in their pursuit of quite different subject matters, a group of social scientists independently of each other appear to converge in a somewhat unorthodox view of certain social phenomena, investigation is in order. The convergence to be examined in this paper is that of the views of Hirschman on economic development, Burton Klein and William Meckling on technological research and development, and Lindblom on policy making in general. These three independent lines of work appear to challenge in remarkably similar ways some widely accepted generalizations about what is variously described in the literature as the process of problem solving and decision making. Before discussing the interrelations of these views, we will give a brief description of each.[1]

Hirschman on Economic Development

A major argument of Hirschman's *Strategy of Economic Development* (1958) is his attack on 'balanced growth' as either a *sine qua non* of development or as a meaningful proximate objective of development policy. His basic defense of *unbalanced growth* is that, at any one point of time, an economy's resources are not to be considered as rigidly fixed in amount, and that more resources or factors of production will actually come into play if development is marked by sectoral imbalances that galvanize

1. Another line of related work is represented in Andrew Gunder Frank's 'conflicting standards' organization theory. It is sufficiently different to fall outside the scope of the present article but sufficiently similar to be of interest to anyone who wishes to explore further the areas of unorthodoxy described here (Frank, 1959; Frank and Cohen, 1959).

private entrepreneurs or public authorities into action. Even if we know exactly what the economy of a country would look like at a higher plateau, he argues, we can reach this plateau more expeditiously through the path of unbalanced growth because of the additional thrusts received by the economy as it gets into positions of imbalance.

Take an economy with two sectors that are interdependent in the sense that each sector provides some inputs to the other and that the income receivers of each sector consume part of the other sector's final output. With *given* rates of capital formation and increase in the labor supply, it is possible to specify at any one time a certain pair of growth rates for both sectors that is optimally efficient from the points of view of resource utilization and consumer satisfaction. This is balanced growth in its widest sense. Unbalanced growth will manifest its comparative initial inefficiency through a variety of symptoms: losses here, excess profits there, and concomitant relative price movements; or, in the absence of the latter, through shortages, bottlenecks, spoilage, and waste. In an open economy, a possible direct repercussion is a balance-of-payment deficit. In other words, sectoral imbalances will induce a variety of sensations – presence of pain or expectation of pleasure – in the economic operators and policy makers, whose reactions should all converge toward increasing output in the lagging sector.

To the extent that the imbalance is thus self-correcting through a variety of market and nonmarket mechanisms, the economy may be propelled forward jerkily, but also more quickly than under conditions of balanced expansion. Admittedly, the process is likely to be more costly in terms of resource utilization, but the imbalances at the same time *call forth* more resources and investment than would otherwise become available. The crucial, but plausible, assumption here is that there is some 'slack' in the economy; and that additional investment, hours of work, productivity, and decision making can be squeezed out of it by the pressure mechanisms set up by imbalances. On the assumption of a given volume of resources and investment, it may be highly irrational not to attempt to come as close as possible to balanced growth; but without these assumptions there is likely to exist such a thing as an 'optimal degree of imbalance'. In other words, within a certain range, the increased economy in the use of given resources that might come with balanced growth is more than offset by *increased resource mobilization* afforded by unbalanced growth.

A simplified geometrical representation of balanced versus unbalanced growth is as follows: let there be two sectors of the economy, such as agriculture and industry, whose outputs are

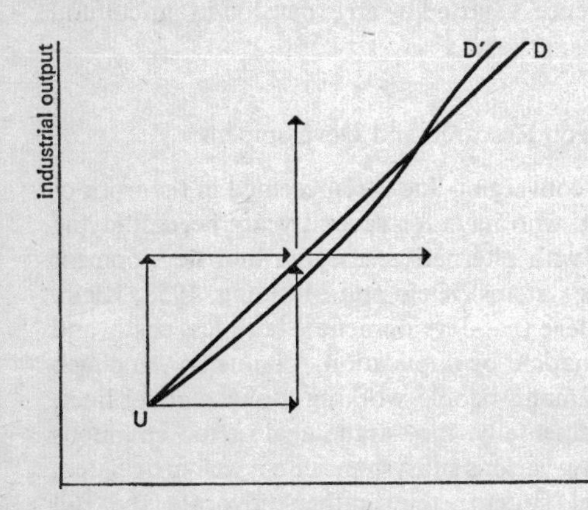

Figure 1 Balanced versus unbalanced growth

measured along the horizontal and vertical axes, respectively. Let point U be the point at which the underdeveloped economy finds itself and D or D′ the goal at which it aims. At this stage, assume certainty and unanimous agreement about this goal. Balanced growth then aspires to a movement along such a line as UD or UD′. At the end of each investment period the economy would find itself producing outputs corresponding to successive points on such lines[2] (see Figure 1). Unbalanced growth means to strike out first in one direction (see arrows) and then, impelled by resulting shortages, balance-of-payments pressures, and other assorted troubles, in the other. Hirschman argues that by traveling along

2. We introduced line UD′ to suggest that balanced growth is not necessarily linear. 'Balance' implies that one knows what the appropriate proportions are at each stage of development, but not necessarily that a constant proportion between two or more absolute rates of growth must be preserved.

this circuitous route, which is likely to be more costly because of the accompanying shortages and excess capacities, the economy may get faster to its goal. Note that there are several varieties of unbalanced growth with varying degrees of pressure. For instance, to start by developing industry is likely to introduce more compelling pressures (because of the resulting food shortages, or, if food is imported, because of the balance-of-payments difficulties) than if the sequence is started by an expansion in agricultural output.

Klein and Meckling on Research and Development

Another apparently converging line is represented in the work of Klein and Meckling, who have for several years been studying military experience with alternative research and development policies for weapons systems (Klein and Meckling, 1958; Klein, 1958, 1960). They allege that development is both less costly and more speedy when marked by duplication, 'confusion', and lack of communication among people working along parallel lines. Perhaps more fundamentally, they argue against too strenuous attempts at integrating various subsystems into a well-articulated, harmonious, general system; they rather advocate the full exploitation of fruitful ideas regardless of their 'fit' to some preconceived pattern of specifications.

Suppose a new airplane engine is to be developed and we know that it ought to have certain minimal performance characteristics with respect to, say, range and speed. A curve such as SS in Figure 2 may represent this requirement. Is anything to be said here in favor of approaching the goal through an unbalanced path, rather than through shooting straight at the target?

The first and perhaps most important point made here by Klein and Meckling is that there is no single point to shoot at, but a great number of acceptable combinations of the two performance characteristics (shown in Figure 2 by the set of all points lying to the northeast of the curve SS). It is perfectly arbitrary for anyone to pick out a point such as S′ as *the* target to shoot at even though this point may be in some sense the expected value of the desired technological advances. The argument then proceeds to show that because of this wide range of acceptable outcomes, and because of the uncertainty as to what is achievable, *any* advance in the north-easterly direction (such as PP′) should be pushed and capitalized on, rather than bent at great effort in the direction of any arbitrarily predetermined target.

The assumption here is that inventions and technical progress follow a 'path of their own' to which we should defer: in other words, instead of getting upset at an early stage of development

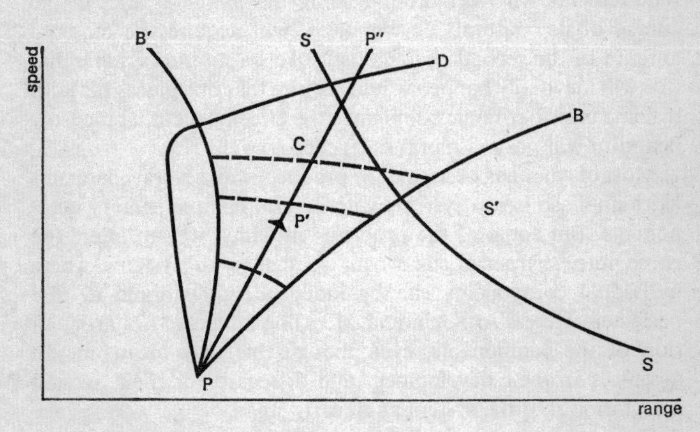

Figure 2 Alternative paths of development of two performance characteristics

with the 'lack of balance' between the two performance specifications (the engine that is being developed is all speed and very little range), we should go on developing it as best we can without reference to point S′. The simplest reason for this is that we may land anyway with a combination of the two characteristics that is acceptable for the purpose at hand: at P″ we have much more speed than we originally bargained for and enough range.

But then there may be other, more interesting reasons why 'a wise and salutary neglect' of the balance between the two performance requirements may be desirable in the earlier stages of research and development. A second possibility is that, as an invention or technological advance matures and is fully articulated, possibilities of adjustment may appear that are not present earlier. In Figure 2 we represent this phenomenon by two boundaries

PB and PB′ that limit the range within which trade-offs between the two characteristics (along the dotted curves) are possible. If these boundaries diverge as shown in our figure, then we should postpone our attempt at trade-offs until we reach the range of greater flexibility (point C).

Third, sometimes the new product that is being developed and which at one stage seemed to be so top-heavy with one of our two requirements will veer around along the path PD and, in the course of its 'natural' development, will acquire the required amount of the second characteristic. To be sure, to assume that this will inevitably happen would require that one places his faith in some basic harmonies, similar to the Greek belief that the truly beautiful will possess moral excellence as well.[3]

Most of what has been said for products with several characteristics applies also to systems with several complementary components. But some of the problems in which we are interested come into sharper focus when we deal with systems where individual components can be independently worked at and perfected. Here also Klein and Meckling advocate full articulation of the components, even though this may mean uneven advances in their development and disregard for their over-all integration into the system at an early stage.

Once again, a principal reason is uncertainty. The final configuration of the system is unknown, and knowledge increases as some of the subsystems become articulated. In the first place, knowledge about the nature of one subsystem increases the number of clues about the desirable features of another, just as it is easier to fit in a piece of a jigsaw puzzle when some of the surrounding pieces are already in place. Second, if two pieces (subsystems) have been worked at independently, it is usually possible to join them together by small adjustments: what is important is to develop the pieces, even though they may not be perfectly adjusted to each other to start with.

Obviously if the subsystems are being perfected fairly independently from one another it is likely that one of them will be fully developed ahead of the others, a situation quite similar to

3. To assume the existence of such basic harmonies may be foolish, but it certainly helps us in making crucial decisions such as choosing a wife or a profession.

that where one sector of the economy races ahead of another. Also it is likely that even if they reach the point of serviceability together, some of them will be 'out of phase' with the others, as in the case of a hi-fi system with an amplifier that is far too good for the loudspeaker.[4]

Lindblom on Policy Making

A third converging line is represented in Lindblom's papers on policy-making processes (1958a–c, 1959, 1961). These papers aspire to fairly large-scale generalizations or to what, in some usages, would be called theory construction; while the points of departure of Hirschman and of Klein and Meckling and two widely different, but still fairly specific, problem-solving contexts. The differences among the studies in this respect make the convergences all the more noteworthy.

Lindblom's point of departure is a denial of the general validity of two assumptions implicit in most of the literature on policy making. The first is that public policy problems can best be solved by attempting to understand them; the second is that there exists sufficient agreement to provide adequate criteria for choosing among possible alternative policies. Although the first is widely accepted – in many circles almost treated as a self-evident truth – it is often false. The second is more often questioned in contemporary social science; yet many of the most common prescriptions for rational problem solving follow only if it is true.

Conventional descriptions of rational decision making identify

4. Klein (1960) gives a straightforward exposition of the logical and empirical differences between development decisions and decisions to maximize the use of existing resources. He again advocates looseness in goal-setting and gradual, oblique, or multiple approaches to the goal. In doing so, Klein now emphasizes the contrast between the decision maker in established production processes who accepts the relatively small uncertainties he faces as a datum, and the development decision maker whose chief purpose is to reduce the huge variance of the initial estimates so that successive investment and production decisions can be made with increasing degrees of confidence. In addition, he argues that it is of rather secondary interest to the developer to achieve an efficient combination of inputs. His main interest is achieving a breakthrough to a new product or to radically improved performance characteristics.

the following aspects: (a) clarification of objectives or values, (b) survey of alternative means of reaching objectives, (c) identification of consequences, including side effects or by-products, of each alternative means, and (d) evaluation of each set of consequences in light of the objectives. However, Lindblom notes, for a number of reasons such a *synoptic* or comprehensive attempt at problem solving is not possible to the degree that clarification of objectives founders on social conflict, that required information is either not available or available only at prohibitive cost, or that the problem is simply too complex for man's finite intellectual capacities. Its complexity may stem from an impossibly large number of alternative policies and their possible repercussions from imponderables in the delineation of objectives even in the absence of social disagreement on them, from a supply of information too large to process in the mind, or from still other causes.

It does not logically follow, Lindblom argues, that when synoptic decision making is extremely difficult it should nevertheless be pursued as far as possible. And he consequently suggests that in many circumstances substantial departures from comprehensive understanding are both inevitable and on specific grounds desirable. For the most part, these departures are familiar; and his exposition of them serves therefore to formalize our perceptions of certain useful problem-solving strategies often mistakenly dismissed as aberrations in rational problem solving.

These strategies, which we shall call 'disjointed incrementalism', are the following:

1. Attempts at understanding are limited to policies that differ only incrementally from existing policy.
2. Instead of simply adjusting means to ends, ends are chosen that are appropriate to available or nearly available means.
3. A relatively small number of means (alternative possible policies) is considered, as follows from 1.
4. Instead of comparing alternative means or policies in the light of postulated ends or objectives, alternative ends or objectives are also compared in the light of postulated means or policies and their consequences.
5. Ends and means are chosen simultaneously; the choice of means does not follow the choice of ends.
6. Ends are indefinitely explored, reconsidered, discovered, rather than relatively fixed.

7. At any given analytical point ('point' refers to any one individual, group, agency, or institution), analysis and policy making are serial or successive; that is, problems are not 'solved' but are repeatedly attacked.
8. Analysis and policy making are remedial – they move *away* from ills rather than *toward* known objectives.
9. At any one analytical point, the analysis of consequences is quite incomplete.
10. Analysis and policy making are socially fragmented; they go on at a very large number of separate points simultaneously.

The most striking characteristic of disjointed incrementalism is (as indicated in 9) that no attempt at comprehensiveness is made; on the contrary, unquestionably important consequences of alternative policies are simply ignored at any given analytical or policy-making point. But Lindblom goes on to argue that through various specific types of partisan mutual adjustment among the large number of individuals and groups among which analysis and policy making is fragmented (see 10), what is ignored at one point in policy making becomes central at another point. Hence, it will often be possible to find a tolerable level of rationality in decision making when the process is viewed as a whole in its social or political context, even if at each individual policy-making point or center analysis remains incomplete. Similarly, errors that would attend over-ambitious attempts at comprehensive understanding are often avoided by the remedial and incremental character of problem solving. And those not avoided can be mopped up or attended to as they appear, because analysis and policy making are serial or successive (as in 7).

While we cannot here review the entire argument, Lindblom tries to show how the specific characteristics of disjointed incrementalism, taken in conjunction with mechanisms for partisan mutual adjustment, meet each of the characteristic difficulties that beset synoptic policy making: value conflicts, information inadequacies, and general complexity beyond man's intellectual capacities. His line of argument shows the influence of pluralist thinkers on political theory, but he departs from their interest in the control of power and rather focuses on the level of rationality required or appropriate for decision making.

Points of Convergence

If they are not already obvious, specific parallels in the works reviewed are easy to illustrate. Compare, for example, an economy that is in a state of imbalance as the result of a sharp but isolated advance of one sector and a weapons system that is out of balance because a subsystem is 'too good' in relation to the capacity of another system. Just as for a sector of the economy, it is possible that a completed subsystem is 'too advanced' only in comparison with some preconceived notion, and that actually its unexpectedly high performance level is quite welcome, either because it improves upon over-all system performance or because it happily compensates for the lag of some other component behind the norms originally set. On the other hand, a component can be 'too advanced' in a real sense; as in a hi-fi set, where the performance of a component depends not only on its capacity but also on inputs from other components. This situation corresponds exactly to that of an economy in structural imbalance. The laggard components turn into bottlenecks for the full utilization of the *avant-garde* component's capacity. Yet even though such a system or economy represents in itself an inefficient utilization of inputs, it may nevertheless be a highly useful configuration if it is conceived as a stage in the development process. For it may be expected that attempts will be made to improve the weaker subsystems or sectors so that the capability of the stronger ones may be fully utilized. In the process, the weaker systems or sectors may be so improved that they become the stronger ones, and the stage thus set for a series of seesaw advances which may carry the overall 'goodness' of our system or economy beyond what might have been achieved by maintaining balance.

For both economy and weapons system we are talking in terms of probabilities. There can be no certainty that with one *avant-garde* subsystem readied the others will dutifully be put in place or improved. The existence of the Maginot Line along the French–German border failed to call forth a corresponding effort along the Belgian frontier to guard against the possibility of a German strategy aimed at circumventing the Line.

This example illustrates an important point: a 'system' or economy is never quite finished. Today's system or economy-in-balance is likely to turn into tomorrow's subsystem or economy-out-of-balance, because of unforeseeable repercussions, newly emerging difficulties, unanticipated counterstrategies, changing tastes or techniques, or whatever other forces with which the system or economy has to deal.[5] But these repercussions, difficulties, and counterstrategies could not possibly be fully visualized in advance. The transportation system consisting of highways, gasoline and repair stations, and automotive vehicles is found incomplete; first because of inadequate accident prevention, and later also because of smog. The new system of defense against infections through antibiotics is suddenly 'out of balance' because of the development of new varieties of drug-resistant micro-organisms. In these cases, it would have been impossible to foresee the imbalance and incompleteness that emerged clearly only after the new system had been in operation for some time.

Once it is understood that a system is never complete or will never stay complete, the case against spending considerable effort on early integration and simultaneous development of subsystems is further strengthened. For if we do achieve early integration and simultaneity, we are much more likely to succumb to the illusion that our system is actually complete in itself and needs no further complements and watchfulness than if we had built it up as a result of seesaw advances and adjustments which do not provide for a natural resting place.[6]

5. It is hardly necessary to mention the similarity, in this respect, between the Maginot Line and some of our present defense systems such as the D.E.W. Line.

6. The examples of the Maginot Line, of automobile traffic, and of antibiotics bring up an additional problem. In the latter two cases, the incompleteness of the system is forcefully brought to our attention through accidents and eye irritation, and through new types of infection. The trouble with some other systems that turn into subsystems is that the mutation may not be so easily detected, or that it may be detected only when it is too late, as was precisely the case with the Maginot Line.

There is real difficulty about the meaning of 'too late'. The imperfections of automobile traffic and antibiotics were discovered too late for the victims of accidents and new-type infections, but not too late, we hope, for the rest of us. The defects of the Maginot Line were discovered too late to save France in 1940, although not too late to win the war against Hitler. This suggests that there may be cases where we cannot afford to do our learning about the imperfections and imbalances of a system through the failures, irritations,

As another specific illustration of convergence, consider the sequence of moves in problem solving as described, on the one hand, in developmental terms by Hirschman, Klein and Meckling, and, on the other hand, in political terms by Lindblom. Recall the picture of desired progress where we wished to move from one fixed point (the present) to another fixed point in a two-dimensional diagram. From existing levels of output in industry and agriculture (or range and speed in aircraft) we wished to move to higher levels for both. Imagine a situation in which two parties with different preferences want to go off in two different directions. Lindblom argues that in this situation the best way to make progress is through 'mutual adjustment', i.e. by a series of moves and countermoves in the course of which a higher plateau can be reached even without prior agreement about the eventual goal. 'Individuals often agree on policies when they cannot agree on ends. Moreover, attempts to agree on ends or values often get in the way of agreements on specific policies' (Lindblom, 1958b, p. 534). Furthermore, it is possible, and even likely, that the value systems of the two parties will move more closely together once an advance that is tolerable to both has been achieved. 'The decision-maker expects to learn about his values from his experiences and he is inclined to think that in the long run policy choices have as great an influence on objectives as objectives have on policy choices' (Lindblom, 1961, p. 309).

Lindblom's reasoning reinforces the others. It parallels Klein and Meckling's emphasis on the inevitability of moving forward through move and countermove, in what appears an arbitrary,

---

and discomforts that are the natural concomitants and signals of the imbalance. Such situations present us with a well-nigh insoluble task, similar to the one which would face a child who had to learn to walk without being permitted to fall. Here the temptation is particularly strong to prepare in advance a perfect theoretical solution. Yet we know from all that has been said that reliance on such a solution would be most likely to bring about the failure one is seeking to avoid. One way of dealing with situations in which we feel we cannot afford to learn the 'hard way' is to develop institutions whose special mission it is to be alert to and to detect existing and developing system imbalances: in a democracy, some institutions of this kind are a free press and an opposition party. For national defense a certain amount of interservice rivalry may serve the same purpose, as each service has a vested interest in pointing out the 'holes' in the other services' systems.

somewhat aimless fashion, rather than Hirschman's stress on the efficiency of such a sequence in squeezing out additional resources. Nevertheless, the idea that unbalanced or seesaw advances of this kind are efficient in some sense is also present. Instead of focusing on the limited supply of decision makers and on the desirability of placing some extra pressures behind investment decisions, Lindblom emphasizes the limited supply of knowledge and the limited area of agreement that exists among various powerholders, and visualizes a series of sequential adjustments as a way to maximize positive action in a society where ignorance, uncertainty, and conflict preclude not only the identification, but even the existence, of any 'best' move.

But we can do better than illustrate parallels. We can explicitly identify the principal points of convergence.

1. The most obvious similarity is that all insist on the rationality and usefulness of certain processes and modes of behavior which are ordinarily considered to be irrational, wasteful, and generally abominable.

2. The three approaches thus have in common an attack on such well-established values as orderliness (see Hirschman's 'model of optimum disorderliness' [1958, p. 80]), balance, and detailed programming; they all agree with Burke that some matters ought to be left to a 'wise and salutary neglect'.[7]

3. They agree that one step ought often to be left to lead to another, and that it is unwise to specify objectives in much detail when the means of attaining them are virtually unknown.

4. All agree further that in rational problem solving, goals will change not only in detail but in a more fundamental sense through experience with a succession of means–ends and ends–means adjustments.

5. All agree that in an important sense a rational problem solver wants what he can get and does not try to get what he wants except after identifying what he wants by examining what he can get.

6. There is also agreement that the exploration of alternative uses of resources can be overdone, and that attempts at

7. An even higher authority might be invoked, namely the Sermon on the Mount: 'Take therefore no thought for the morrow, for the morrow will take care of the things of itself.'

introducing explicitly certain maximizing techniques (trade-offs among inputs or among outputs, cost-benefit calculations) and coordinating techniques will be ineffective and quite possibly harmful in some situations. In a sense more fundamental than is implied by theories stressing the cost of information, the pursuit of certain activities that are usually held to be the very essence of 'economizing' can at times be decidedly uneconomical.

7. One reason for this is the following: for successful problem solving, all agree it is most important that arrangements exist through which decision makers are sensitized and react promptly to newly emerging problems, imbalances, and difficulties; this essential ability to react and to improvise readily and imaginatively can be stultified by an undue preoccupation with, and consequent pretense at, advance elimination of these problems and difficulties through 'integrated planning'.

8. Similarly, attempts at foresight can be misplaced; they will often result in complicating the problem through mistaken diagnoses and ideologies. Since man has quite limited capacities to solve problems and particularly to foresee the shape of future problems, the much maligned 'hard way' of learning by experiencing the problems at close range may often be the most expeditious and least expensive way to a solution.

9. Thus we have here theories of successive decision making; denying the possibility of determining the sequence *ex ante*, relying on the clues that appear in the course of the sequence, and concentrating on identification of these clues.

10. All count on the usefulness for problem solving of subtle social processes not necessarily consciously directed at an identified social problem. Processes of mutual adjustment of participants are capable of achieving a kind of coordination not necessarily centrally envisaged prior to its achievement, or centrally managed.

11. At least Hirschman and Lindblom see in political adjustment and strife analogues to self-interested yet socially useful adjustment in the market.

12. All question such values as 'foresight', 'central direction', 'integrated overview', but not in order to advocate *laissez faire* or to inveigh against expanded activities of the state in economic or other fields. They are in fact typically concerned with decision-

making and problem-solving activities carried on by the state. In their positive aspects they describe how these activities are 'really' taking place as compared to commonly held images; and in so far as they are normative they advocate a modification of these images, in the belief that a clearer appreciation and perception of institutions and attitudes helpful to problem-solving activities will result.

Although many of these propositions are familiar, they are often denied in explicit accounts of rational decision making; and at least some of them challenge familiar contrary beliefs. Either the convergences are an unfortunate accident, or decision-making theory has underplayed the degree to which 'common sense' rational problem-solving procedures have to be modified or abandoned. Account must be taken of man's inertia, limited capacities, costs of decision making, and other obstacles to problem solving, including uncertainty, which is the only one of the complicating elements that has been given sustained attention. And most investigations of uncertainty have been within the narrow competence of statistical theory.

Points of Difference

These similarities in approach, with their widely different origins, structures, and fields of application, are even better understood if their remaining points of difference are identified.

The basic justification for rejecting traditional precepts of rationality, planning, and balance is somewhat different for the three approaches here examined. For Lindblom it is *complexity*, i.e. man's inability to *comprehend* the present interrelatedness and future repercussions of certain social processes and decisions, as well as imperfect knowledge and *value conflicts*. For Klein and Meckling it is almost entirely *future uncertainty*, i.e. man's inability to *foresee* the shape of technological breakthroughs, or the desirability of letting oneself be guided by these breakthroughs if and when they occur, instead of following a predetermined sequence. For Hirschman it is the difficulty of mobilizing potentially available resources and decision-making activity itself; the *inadequacy of incentives* to problem solving, or, conversely, the need for *inducements* to decision making.

Although Klein's and Meckling's concern with future uncertainty could formally be viewed as a special case of Lindblom's problem of inadequate information, their treatment of the research and development problem is different enough from Lindblom's treatment of information inadequacies to argue against its being so viewed. Hirschman's concern with inducements to problem-solving activity is quite different from either Lindblom's or Klein's and Meckling's concern with limits on cognitive faculties. He argues not that men lack knowledge and capacity to solve problems in an absolute sense, but that there is always some unutilized problem-solving capacity that can be called forth through a variety of inducement mechanisms and pacing devices. These different reasons for supporting the same conclusions make the conclusions more rather than less persuasive, for the reasons supplement rather than invalidate each other.

That they are complementary reasons is, of course, indicated by the overlap of the Lindblom and the Klein–Meckling approaches on the problem of imperfect information, and by some Hirschmanlike concern for research and development *incentives* in the Klein–Meckling study. It is also true that Hirschman develops as a secondary theme the difficulties of ignorance and uncertainty in economic development. For instance, his partiality toward 'development via social overhead capital shortage' is based in part on the position that shortages and bottlenecks remove uncertainty about the direction of needed overhead investments. Similarly, he emphasizes the importance of unforeseen or loose complementarity repercussions, such as 'entrained wants' that arise in the course of development, and asserts that imports are helpful in inducing domestic production because they remove previous doubts about the existence of a market.

From the differences in the main thrust of the respective arguments, certain other major differences emerge, differences which do not deny the convergences, but which, on the other hand, ought not to be submerged by them. For example, Hirschman's argument that a very heavy reliance on central planning will often be inappropriate for underdeveloped countries looks superficially parallel to Lindblom's argument that *partisan* mutual adjustment can sometimes achieve efficiencies that could not be achieved through over-ambitious attempts at central omniscience and control. Yet on closer scrutiny, Hirschman's cautions about centralism only secondarily refer to the *general* difficulties of managing complex affairs that strain man's incentives and intellectual capacities. Instead he argues that a conventional, centrally planned attempt to define and achieve a balance among many varied lines of development will be less helpful than a similarly central attempt to estimate and manage the critical linkages through which economic growth is forced or induced.[8]

Hirschman's explicitly declared view of decision making for economic development is almost entirely one of central planning, or at least problem solving by persons – such as planning board managers or officials of international agencies – who assume some general responsibility toward the economy as a whole, and whose point of view is therefore that of a central planner. Hirschman's policy maker or operator is, with only a few exceptions, such a person or official; and Hirschman's prescriptions are always addressed to such a person. By contrast, Lindblom's policy maker is typically a partisan, often acknowledging no responsibility to his society as a whole, frankly pursuing his own segmental interests; and this is a kind of policy maker for whom Hirschman, despite his between-the-lines endorsement of him, makes no explicit place in his formulation of the development process.

A further important point of difference between Hirschman and Lindblom appears to lie in Hirschman's emphasis on discovering and utilizing the side-effects and repercussions of development decisions, as compared to Lindblom's readiness to recommend at any given 'point' neglect of such repercussions. It is indeed a major thesis of Hirschman that analysis of a prospective investment project should above all try to evaluate its effect on further development decisions instead of conventionally concentrating on its own prospective output and productivity. Specifically, every decision should be analysed to discover its possible 'linkages' with other decisions that might follow it. For example, a prospective decision to encourage the importation of some consumer goods, such as radios, should consider not simply

8. This argument against the attempt at balanced growth is quite different from Hirschman's other argument that balance in growth is not desirable even if achieved.

the economy's need for these goods but the probability that their importation will in time lead to a decision by domestic investors to assemble them locally, as well as the 'linkage effects' of such assembly operations on further domestic production decisions.

Hirschman's book (1958) is both an attempt to uncover such linkages and a prescription that developers seek to uncover them in every possible case. Lindblom suggests that this kind of by-product, the indirect consequences of a decision that flow from the decision's effect on still other decision makers will often escape the analyst in any case; hence he should not try to always anticipate and understand it, but instead should deal with it through subsequent steps in policy making, if and when it emerges as a problem. Since, as Lindblom sees it, policy making is not only *remedial* and *serial* but also *fragmented*, both intentionally and accidentally neglected consequences of chosen policies will often be attended to either as a remedial next step of the original policy makers or by some other policy-making group whose interests are affected. Hence policy as a complex social or political process rises to a higher level of comprehensiveness and rationality than is achieved by any one policy maker at any one move in the process.

The contrast between Hirschman and Lindblom on this point can be overdrawn, however. For one thing, Hirschman feels that calculations which purport to give greater rationality to investment planning may often interfere with development, because they typically do not and cannot take the 'linkages' into account; whereas more rough-and-ready methods may be at least based on hunches about such linkages. Second, Hirschman's practical advice to policy makers is similar to Lindblom's when he tells them to go ahead with unintegrated and unbalanced projects on the ground that, in an interdependent economy, progress in some sectors will serve to unmask the others as laggards and will thereby bring new pressures toward improvement. In his general prescription, more implicit than explicit, that development planners try to move the economy wherever it can be moved, that is, seize on readiness to act wherever it can be found, Hirschman is endorsing Lindblom's suggestion that many consequences can best be dealt with only as they actually show themselves.

As a further point of difference, it is implicit in what has been said in the preceding paragraphs that Hirschman's thinking about secondary effects is preoccupied with possible bonuses to be exploited, Lindblom's with possible losses to be minimized. Again, the difference is easy to overstate: Hirschman too is at times concerned with possible losses, even if Lindblom has not explored at all the possibility of bonuses. Hirschman, however, relies on correct diagnosis of linkages for protection from damaging side effects; and his position is therefore parallel to his position on exploiting bonus effects. Only secondarily (1958, pp. 208 ff.) does he count on Lindblom's remedial, serial, and fragmented kind of process for minimizing losses.

Concluding Remarks

As Hirschman would now give uncertainty, complexity, and value conflict a more central place in justifying his conclusions on economic development policy, so also Lindblom's and Klein's and Meckling's analyses could be strengthened by taking into account the fact that the policies they defend could also be justified because they permit mobilization of resources and energies that could not be activated otherwise. Perhaps these latter analyses could go beyond the statement that the processes of research and development and of policy making are of necessity piecemeal, successive, fragmented, and disjointed; they could try to define typical sequences and their characteristics, similar to Hirschman's 'permissive' and 'compulsive' sequences. Once the intellectual taboo and wholesale condemnation are lifted from some of the policies Klein, Meckling and Lindblom defend, it becomes desirable to have a closer look at the heretofore incriminated processes and to rank them from various points of view. It is useful to ask questions such as the following: as long as we know that a system is going to be out of balance anyway when the subsystems develop, what type of imbalance is most likely to be self-correcting? An answer to this question could affect the desirable distribution and emphasis of the research and development effort. Detailed descriptions of types of incremental meandering would also be interesting; perhaps this would more clearly differentiate between a sequence that leads to reform and another that leads to revolution.

One problem deserves to be mentioned again. The processes of economic development, research and development, and policy making must all rely on successive decision making because they all break new, uncertain ground. Therefore these processes must let themselves be guided by the clues that appear en route. Snags, difficulties, and tensions cannot be avoided, but must on the contrary be utilized to propel the process further. The trouble is that the difficulties are not only 'little helpers', but may also start processes of disintegration and demoralization. An intersectoral imbalance sets up a race between the catching-up, forward movement of the lagging sector and the retrogression of the advanced one. The greater the pressure toward remedial positive action, the greater is the risk if this action does not take place. There is a corresponding situation in systems development. The more a system is out of balance, the greater will presumably be the pressure to do something about it, but also the more useless is the system should no action be forthcoming.

All three approaches therefore have one further characteristic in common: they can be overdone. There are limits to 'imbalance' in economic development, to 'lack of integration' in research and development, to 'fragmentation' in policy making which would be dangerous to pass. And it is clearly impossible to specify in advance the optimal doses of these various policies under different circumstances. The art of promoting economic development, research and development, and constructive policy making in general consists, then, in acquiring a feeling for these doses.

This art, it is submitted by the theories here reviewed, will be mastered far better once the false ideals of 'balance', 'coordination', and 'comprehensive overview' have lost our total and unquestioning intellectual allegiance.

References

Frank, A. G. (1959), 'Goal ambiguity and conflicting standards: an approach to the study of organization', *Human Organization*, vol. 17, pp. 8–13.

Frank, A. G., and Cohen, R. (1959), *Conflicting Standards and Selective Enforcement in Social Organization and Social Change: A Cross Cultural Test*, Paper read at the American Anthropological Association Meeting, Mexico City, December.

HIRSCHMAN, A. O. (1958), *The Strategy of Economic Development*, Yale University Press.

KLEIN, B. (1958), 'A radical proposal for R and D', *Fortune*, May, p. 112.

KLEIN, B. (1960), *The Decision-making Problem in Development*, Paper no. P-1916, The RAND Corporation, Santa Monica, California, 19 February.

KLEIN, B., and MECKLING, W. (1958), 'Application of operations research to development decisions', *Operations Research*, vol. 6, pp. 352–63.

LINDBLOM, C. E. (1958a), 'Policy analysis', *American Economic Review*, vol. 48, pp. 298–312.

LINDBLOM, C. E. (1958b), 'Tinbergen on policy making', *Journal of Political Economy*, vol. 66, pp. 531–8.

LINDBLOM, C. E. (1958c), 'The handling of norms in policy analysis', in M. Abramovitz (ed.), *Allocation of Economic Resources*, Stanford University Press, pp. 160–79.

LINDBLOM, C. E. (1959), 'The science of "muddling through"', *Public Administration Review*, vol. 19, pp. 79–88.

LINDBLOM, C. E. (1961), 'Decision making in taxation and expenditure', in Universities-National Bureau of Economic Research, *Public Finances: Needs, Sources, and Utilization*, Princeton University Press.

18 M. Ways

The Road to 1977

M. Ways, 'The road to 1977', *Fortune* (January 1967), pp. 93–5, 194–7, Time Inc.

The youth pictured opposite [see original article for illustration] is a student at one of the better big-city high schools. By 1977, college and graduate school behind him, he will have become a full-fledged member of American society, contributing his trained skills and using his informed values to shape its future. His starting point, 1977, is being shaped by our decisions now. What could we tell him about changes in the quality of our society in the next ten years? What can we predict and what do we intend?

The most confident predictions we can make contain the word 'more' – more automobiles, more color-television sets, more sirloin steaks, more medicines and art museums, more mobility, more opportunity, more variety. Complementary statements look forward to 'less'. The proportion of Americans below 'the poverty line' will continue its long decline. We can be sure that between now and 1977 the fatal or crippling impact of some diseases will be lessened. We intend to have less polluted rivers, less disorganized cities, fewer high-school dropouts – and we can predict some success. In short, 1967 can say that many trends and/or plans now in train will by 1977 be further advanced.

The prospect of 'more' is based not only on the continuing explosion of new technologies but on a broad movement discerned by Alexis de Tocqueville. Examining the structure, environment, and national character of American society a hundred and thirty years ago, he saw in it the pattern of democratic diffusion. His prediction, that in the U.S. the many would demand what in other lands was reserved for the few, has been validated not only by mass prosperity but by universal suffrage, mass literacy, and, in our day, by mass higher education and a start toward mass intellectual employment. One-man-one-vote reapportionment and the civil-rights movement attest that democratic diffusion has not lost its vigor.

This great pattern has within the last few years been joined by another, which is the subject of this article – a new style of private and public planning, problem solving, and choosing. This new style promises to add a missing ingredient to the quality of American life.

Glitter in a Swamp

For all our success in fulfilling human needs, we have all along been haunted and even demoralized by the thought that we, like Poor Richard's friend, have paid too much for our whistle. The unprecedented pace and breadth of change revived in more disquieting form an old question, what does it all mean? And introduced a new social question, where is it all headed? Change, the product mainly of individual decisions, had a haphazard look about it in the aggregate. Our environment came to seem an archipelago of successes glittering in a swamp of unintended consequences. Anyone could look at the cities, sniff their air, understand the alienation of their poor, and say it was a world *he* never made – or, at least, never meant to make. Specialization of knowledge and work required large and complex organizations; these raised fears that individuality would be attenuated, that an 'organization man', bland and malleable, might replace his admirably hard-nosed ancestor. This feared sacrifice of individuality, repugnant on democratic and moral grounds, would not even be compensated for by an augmented sense of social warmth or common purpose. We were moving, obviously, but weren't we adrift?

Such gnawing doubts about progress made the exigent past – the New England village or the prairie farm – seem attractive in retrospect to many Americans. Meanwhile, the doubts and insecurities haunting a changing world led – in the U.S. and abroad – to the rise of ideologies that promised to join change with social cohesion, progress with a sense of purpose and direction, material abundance with the moral goal of social justice. Emanating from several different points on the political compass, these were enticing promises to an uneasy world; but they turned out on all counts (including the practical) to be empty. Now the ideologies and the imitative reactions they provoked are dying in all but the more backward and frustrated parts of the world. In the U.S. they linger as components in the political positions self-described as 'liberal' and 'conservative', each a feeble and primitive effort to find a general way of dealing with an immense new fact of life, radical change.

Fortunately, however, the decline of ideology and its two pale heirs does not return the U.S. to the haphazard condition that originally encouraged ideology's growth. For the new ways of dealing with change can themselves generate in the public mind a sense of direction, of intelligent, effective choice, a sense that our evolving capabilities and our evolving values will move hand in hand.

The further advance of this new style is the most significant prediction that can be made about the next ten years. By 1977 this new way of dealing with the future will be recognized at home and abroad as a salient American characteristic. Compared to this development, the argument between the liberals and the conservatives, while it will retain a certain atavistic fascination, will come to seem about as relevant to the main proceedings as a fistfight in the grandstand during a tense innings of a World Series game.

The new style of dealing with the future has no accepted, inclusive name, but the names of its more highly developed techniques have become familiar in the last ten years to most businessmen, government officials, military officers, scientists, and technicians. The techniques themselves, which are apt to be called 'systems analysis' or 'systems planning', are now widely used both with and without the help of computers. 'Cost-benefit' or 'cost-effectiveness' analysis is a major ingredient of the new techniques; this involves ways of arraying ends and means so that decision makers have clearer ideas of the choices open to them and better ways of measuring results against both expectations and objectives.

Among characteristics of the new pattern are these:

1. A more open and deliberate attention to the selection of ends toward which planned action is directed, and an effort to improve planning by sharpening the definition of ends.

2. A more systematic advance comparison of means by criteria derived from the ends selected.
3. A more candid and effective assessment of results, usually including a system of keeping track of progress toward interim goals. Along with this goes a 'market-like' sensitivity to changing values and evolving ends.
4. An effort, often intellectually strenuous, to mobilize science and other specialized knowledge into a flexible framework of information and decision so that specific responsibilities can be assigned to the points of greatest competence.
5. An emphasis on information, prediction, and persuasion, rather than on coercive or authoritarian power, as the main agents of coordinating the separate elements of an effort.
6. An increased capability of predicting the combined effect of several lines of simultaneous action on one another; this can modify policy so as to reduce unwanted consequences or it can generate other *lines* of action to correct or compensate for such predicted consequences.

Instead of Cut-and-Try

The history of the new style is well known. From scattered beginnings as 'operations analysis' in World War II, it coalesced as an institution in the postwar Rand Corp., which did analytical work in the systematic comparison of weapons for the Air Force. Through the fifties, this mode of presenting alternatives to decision makers had an increasing influence on the Defense Department. Beginning in 1961, Secretary Robert S. McNamara restructured the whole work of the department in this style, programing all planning and procurement around missions or objectives that cut across the boundaries of the three services and extended beyond the confines of annual budgets. (For purposes of congressional appropriations, the programs are translated into conventional annual budget terms by means of 'crosswalks'.) This way of administering the Defense Department proved so successful that in 1965 President Johnson directed that PPBS (planning–programing–budgeting system) be introduced into all departments of the federal government. The new techniques have been helpful in foreign-aid planning and in defining Peace Corps missions. They are beginning to be recognized as the greatest advance in the art of government since the introduction nearly a hundred years ago of a civil service based upon competence. The new style, indeed, corrects an old defect of bureaucratic organization, which was at its best when performing routine tasks, at its worst in innovating and generating forward motion.

Business schools and corporate planners began developing the new technique at about the same time as the Rand Corp.'s early work for Defense. Since then the efficiency of the new methods has been brilliantly demonstrated in the marketplace. As Arjay Miller, president of Ford Motor Co., has put it: 'Hunches and cut-and-try methods are giving way to the systems-analysis approach, a whole new way of perceiving problems and testing in advance the consequences of alternative actions to solve those problems. Computers and other technical devices, including mathematical models, have extended greatly our ability to understand and cope with the complex problems we face in today's world.'

One of the problems of today's world is 'the problem of bigness'. The new style can deal with that by distributing to a larger and larger proportion of the population responsibility for the decisions that shape the future. It can also inculcate a common style of action among business managers, government officials, and university professors; already, more and more people are circulating freely through all three of these formerly walled-off worlds. By mobilizing specialized and value-free science to work on practical problems, the new pattern can help restore the community of scientists and scholars and build an organized link between science and value.

The new style is by no means confined to the world of Big Government, Big Business, and Big Education; at all levels, it's 'in the air'. This young man's [consult original source for illustration] main problem is choosing a college; and, like most of his peers, he has been engaged in a complex strategy of college entrance. He has had to think about career objectives, compare colleges in that light, reconsider them in the light of his tested performance, seek advice (less expert than it is thought to be) from guidance counselors and manuals – all this a far cry from yesterday's daydreamed wish, 'It would be nice to go to Yale'.

Neither for him nor for his elders will the road from 1967 to 1977 be a primrose path. Nothing in the new style offers a guarantee that blunders and frustrations will be reduced (what could?). But for this society and for many members of it, the next ten years will be less like what the mathematicians call a 'random walk'. Therein lies a significance even greater than the new style's efficiency. By trying to get a better grip on the future – but not strangling it with iron utopianism – U.S. society is moving up to a new level of moral and intellectual challenge.

Futurists at Work

Among the intellectual leaders of society a new group, 'the futurists', has come into view. Of four we select as examples, only Henry S. Rowen, new head of the Rand Corp., is heavily engaged with systems planning in the strict sense. Bertrand de Jouvenel is both a pioneer of the new planning and a link between futurists in the U.S. and Europe. He has tried to stimulate among intellectuals a habit of not 'taking the future for granted'.

In similar spirit, the American Academy of Arts and Sciences has set up a group to study the year 1976. Headed by Stephen R. Graubard, editor of *Daedalus*, this group is not seeking a consensus on agreed goals. Rather, it is trying to stir up ideas for comparison and debate in such fields as urban problems, education, health, industry and technology, recreation and the arts.

Daniel Bell is chairman of a group that is even more ambitious in its scope, the Commission on the Year 2000, which is also sponsored by the American Academy. No substantive proposal that Bell's project, or Graubard's, will eventually formulate could be more significant than the *way* these groups conceive their tasks and go about their work. Proceedings (*Daedalus*, Summer, 1967) of the Year 2000 Commission's meetings may serve as source material for historians a century hence when they come to deal with a main theme of the years 1950–2000, the new planning style.

Bell, in a preliminary statement, set the stage for the commission's work: 'Not only is there the awareness that we live, in the trite phrase, in an era of social change, but we begin to realize that it is possible to direct some of this change consciously, that we need to consider the anticipated consequences of change, and we need to seek to control some of the anticipated effects. More than that, we wish to specify alternative consequences of change in order to widen the area of choice, for that normative commitment underlies any humanistic approach to social policy.'

When they met, the commission's members had a hard time deciding what kind of statement about the year 2000 they wished to make. Should it be mainly prediction, based upon the extrapolation of existing trends? Or should they propose what they believed would improve the quality of life by the year 2000? As discussion proceeded, the interaction between prediction and values became plain. At one point Fred C. Iklé, an M.I.T. political scientist, put the matter very simply: 'Let me illustrate by using two extremes. If we had a perfect art of prediction and we knew what the year 2000 would be like, I don't think we would be here. If we had no values we wouldn't be here because we wouldn't care.' Lawrence K. Frank, a retired foundation executive, worried lest 'too exact forecasts' rob people of the feeling of openness 'necessary for a free society'. But Harvey S. Perloff, director of regional studies of Resources for the Future, pointed out how prediction can increase a sense of opportunity.

Public understanding of the anticipated growth of the economy, he said, might make people think about goals they otherwise would ignore. 'If we are not doing the things we can afford to do,' he said, 'and if people know it, they are likely to make different value choices.' Yet all of the group seemed aware of economic limits on the total range of choice. As Martin Shubik of Yale expressed it: 'It is all very lovely to talk about worlds of abundance; but it is just nonsense.' Even in some future society rich enough to spend $100,000 *per capita* for health, education, and welfare, 'There is still that allocation problem; we have to decide how to do it.'

Parameters of Progress

The sense of working in high tension between mounting opportunity and predictable constraints also pervades a two-year study made by the National Planning Association and reported in a recent book, *Goals, Priorities and Dollars – the Next Decade* (Free Press and Macmillan, 1966), by Leonard A. Lecht. This is not one of those doomful prophecies that we are recklessly running out of mineral resources; neither is it an exuberant pep talk for a faster pace of progress. Rather, its concern for the 'fit' of particular plans into some limited program illustrates why the new planners are so intent on selecting the best goals and pursuing them by the most effective means. The message is: we can't do everything, but we can do a lot.

The terminal year of Lecht's projection is 1975. He constructs parameters of progress by estimating to 1975 the gross national product, the population, the civilian labor force. Assuming a high growth rate, he projects a 1975 G.N.P. of $981 billion (in 1962 dollars). He then takes sixteen 'goal areas' including consumer expenditures and savings, private plant and equipment, national defense, housing, urban development, health, education, social welfare, space and transportation. He then projects their costs over the same period. Cost increases are of two kinds: 'preempted bench marks' (which derive from our existing standards) and 'aspirations'. In education, for instance, preempted bench marks can be calculated by taking the costs of meeting existing standards of quality and then adding the cost of necessary expansion – e.g. because we will have an increased number of students; such bench marks will require nearly half of the whole increase in G.N.P. from the actual level of 1962 to the estimated level of 1975. The rest would be available for 'aspiration standards'.

These are goals of improvement, in some cases stated in public policy; in others, such as private plant and equipment, derived from knowledgeable projections of demand. The total net cost of all these goals in the year 1975 comes to $1271 billion (in 1962 dollars), or nearly 15 per cent more than estimated G.N.P. For the 1962–75 span Lecht projects the following percentage increases in costs if we tried to meet our aspirations: consumer expenditures, 85 per cent; national defense, 31 per cent; private plant and equipment, 210 per cent; health, 164 per cent; education, 176 per cent; social welfare, 142 per cent; housing 111 per cent; urban development, 102 per cent; transportation, 113 per cent; research and development, 131 per cent; space, 183 per cent. Note that in all the categories except consumer expenditures and national defense 1975 costs of the aspirations standards would more than double the actual expenditures of 1962; private plant and equipment would more than triple the 1962 figure. In other words, there will still be 'that allocation problem'.

If the U.S. does not use its predictive faculty to measure the feasibility of its aspirations, then it may run into such unintended consequences as shortages, inflation, and political disruptions. Or conversely, by failing to predict the magnitude of future problems and by underrating the society's capability to handle them, the present generation may hand over to the next a mess of avoidable crisis more formidable than those with which we now desperately and belatedly deal: civil rights, air and stream pollution, etc.

The Jeremiahs versus the Savonarolas

Many of the government's social programs are connected with the presence in the U.S. of more than 30 million men, women, and children living below the 'poverty line'. As a proportion of the total population, this group has been receding for decades from the point where the poor were the majority. From 1959 to 1965 the group below the poverty line, as defined by the Social Security Administration, dropped from 22 per cent to 17 per cent of the population, and by 1977 it is expected to drop below 12 per cent. Most of the long improvement is attributable to general American progress, the Tocquevillian diffusion, assisted to a substantial degree by governmental action.

Gratifying as this rate of progress is by some standards, there is plenty of evidence of a large and growing American consensus that it ought to move faster. The very fact that the poverty group is shrinking makes the problem seem more manageable. At the same time, there is recognition that as the group becomes smaller the suffering of its individual members may increase. Many citizens believe public relief, as now set up, discourages the poor from helping themselves to rise.

From these trends of conditions and opinion one can predict that in the next ten years public interest in better methods of handling 'welfare' will increase. The interesting question is how the discussion will be organized. Will it be a class-and-ideological issue, dominated by such fatuous slogans as 'human rights *versus* property rights'? Or can this problem, too, be shifted from the politics of issues to the politics of problems? There are indications that it can. After years of sterile and bilious debate over generalities, the air is suddenly filled with quite detailed proposals, such as a guaranteed annual income and a negative income tax, for handling the poor. The public is recognizing the need for more facts, for more careful definition of what we are trying to do, and for a sharper comparison of alternative methods. There is a good chance that by 1977 the quality of the debate over welfare will have advanced markedly beyond the level of futile filibusters involving the country-club Jeremiahs versus the faculty-club Savonarolas.

The Public Market

What, meanwhile, of the part of American society which produces the wherewithal that makes the mare go? If, as is argued here, the new style of planning and problem-solving will strengthen the hand of government, won't that improvement be bad for 'the other sector' – business?

It need not and, probably, it will not be. In the first place, business, for reasons internal to itself and independent of government, had reached a point where it badly needed a new way of organizing its own movement into the future. Technological complexity and the accelerating pace of change, both largely generated within the business world, had made the world unsafe for corporations using managerial methods developed to fit a more static situation (see 'Tomorrow's management', *Fortune*, 1 July 1966). Whatever the troglodytes of the Antitrust Division of the Justice Department may say to the contrary, competition in U.S. business is becoming more lively year by year: more exports into markets where U.S. corporations have small shares; more imports into the U.S.; more new products and processes; more potential consumer 'swing' among the rising proportion of discretionary goods and services. In those circumstances, today's milestone of corporate success could become tomorrow's tombstone if business had not in recent years discovered a mode of management more internally flexible, more attuned to deal with technological complexity, and more oriented toward the future.

In the second place, the simultaneous advance of the new style in both business and government is going to make it easier for them to deal with each other. For the next several decades, at least, dealings between business and government on all levels will probably increase, and it is important to understand why.

The true trend in the government–business relationship has been obscured by a fashion that decrees we must speak of a 'public sector' and a 'private sector'. In the U.S. the rough-and-ready way of dividing them is to add federal, state, and local government expenditures and express the total as a percentage of G.N.P. For 1965 that gives a 'public sector' of 27·3 per cent; the other 72·7 per cent is deemed 'private sector'. Whenever the 'public sector' percentage rises in this crude calculation, some citizens react with horror, others with resignation, and still others with joy, in the common belief that what is going on here is 'creeping socialism'. This conveniently overlooks socialism's central characteristic: public ownership of the means of production. In the U.S. nothing approaching 27·3 per cent of goods and services is produced by federal, state, and local governments. Of the total $186 billion spent by them in 1965, only $68 billion, or about 10 per cent of G.N.P., went for 'general government' expenditures. More than half the rest was used by governments to purchase goods and services produced privately – weapons, space capsules, buildings, vehicles, research, etc. The old demands that government nationalize railroads, coal mines, shipping, shipbuilding, arms-making have in the last thirty years subsided from a roar to a whisper. Instead, governments as mass purchasing agents have operated increasingly in competitive markets.

It is this trend, not 'creeping socialism', that can be expected to increase for some years. Individuals cannot buy cleaner rivers or cleaner urban air for themselves. Government agencies are going to make these purchases, but corporations are going to sell the equipment that launders streams and atmosphere. A tremendous opportunity for private business expansion lies in this area, which J. Herbert Hollomon, Assistant Secretary of Commerce for Science and Technology, calls 'the public market'. Hollomon would like to see, for example, a profit-making organization running a chain of junior colleges under contracts with the communities in which they are located. The private company might develop a curriculum, pay what the market required to recruit teachers of quality, and sell the package in competition with other junior-college chains. Hollomon is intensely interested in research-and-development work on *de novo* cities now going on in several companies (Litton Industries, for instance). 'I see no reason why you can't have privately financed entrepreneurial cities of 400,000 people,' he says. 'The public needs are going to be met. The question is whether or not we are going to meet them intelligently.'

Moreover, the concept of 'the public market' holds huge long-range opportunities for private business to 'creep' into service fields traditionally dominated by government. Control of traffic flow, public-library functions, street cleaning, refuse disposal, are obvious possibilities.

Business–government relations in such expanding public markets will certainly call for an approach very different from the way municipalities now let contracts for the construction of a jail or the purchase of a truck fleet. Already there is talk of forming a Comsat-type corporation to handle the complex relations of private businesses and government agencies in urban-redevelopment projects.

The new federal Department of Housing and Urban Development is not panting to spend money as fast as possible. Its planners do not assume they have ready-made answers to the complex problems of the cities. They hope in the next three or four years to mobilize intellectual resources for an attack on those problems. The 'demonstration cities' plan, that $900-million 'drop in the bucket', will give them a chance to test the better programs developed by municipalities. By the early or middle seventies H.U.D. planners hope that the U.S. will be able to mount, through private and public efforts, a more massive drive to renew the cities. If such a drive succeeds, systematic planning will be an important part of it.

How the Larchmonts may Survive

A weak link can be identified in many new programs. State and local governments, though the quality of some has improved markedly, are generally lagging behind the onrush of their new problems. For decades, there has been a grave danger that the U.S. would become a politically centralized nation because of the inability of state and local governments to cope with the complex problems of change. The new style of planning and problem-solving offers substantial hope for resuscitating state and local government.

To take one category of trouble, consider the confusion created by thousands of local governments within the great and growing metropolitan areas. Up until a few years ago most students of U.S. politics believed that these local governments would have to disappear and be replaced by some consolidated form of general metropolitan government. Sensible as this conclusion might seem on paper, it presented hideous difficulties as a practical political program. Larchmont – figuratively speaking – would fight on the hills and on the beaches before it was consolidated with the Bronx – and the Bronx, indeed, might not rejoice at being engulfed (or engolfed) by all the Larchmonts. While many politicians understood the need for consolidated metropolitan government, few were willing to risk their careers in so unpopular a cause. It could be predicted in 1947 that this quiet and deep-seated stalemate would not be broken for decades. Central cities would continue their dependence on the federal government and metropolitan areas would remain helpless to deal with problems cutting across town lines.

But now a third alternative appears. Students of government have come to believe that many of the existing local political units can be effectively coordinated without losing their cherished autonomy. The metropolitan-area issue is being gradually analysed – broken up – into a set of problems. Since it is obvious that Larchmont or the Bronx or even New York City, acting alone, cannot buy itself clean air there will have to be some co-ordinating regional anti-pollution authority. Similarly, metropolitan-area traffic problems cut across local boundaries, but no 'traffic czar' in a central, megalopolitan hall is needed to deal with them – nor would he be capable of doing so. What it takes is systems planning of traffic flow, a sensitive and coherent network of information and of local decisions that 'fit' one another. Some measure of coercion, probably from state capitals, will be needed to bring such local coordination into being – but this will be much less unpalatable to local communities than a general metropolitan government would be.

Brains Instead of Muscle

In law enforcement the chaos created by the multiplicity of police forces – local, county, and state – prompted many citizens to wish for a national police force. They said that was the only way to cope with organized crime and with the increased mobility that the automobile and the telephone gave to criminals. But as this problem is examined more closely it becomes clear that there are a great many steps that might be taken short of nationalized law enforcement. For instance, exchange of information and other forms of cooperation between police forces have been haphazard, tardy, and woefully incomplete. Now computerized communications systems are aiding police in Los Angeles, Chicago, Detroit, and other cities. In police work as in business, better information networks may make decentralization possible by providing a mode of coordination more effective than a monolithic power structure.

In this direction lies American society's response to fears, voiced by many conservatives, that recent U.S. Supreme Court decisions protecting the rights of persons accused of crime have made effective enforcement of criminal law impossible. By overhauling the antiquated techniques of police management we may be able to more than offset the police problems arising from the Court's insistence on more careful concern for individual rights. What law enforcement needs, perhaps, is not more 'power' but more brains.

How Computers Aid a Moral Advance

Turning to a very different kind of police work, one can see in the conduct of the Vietnam war how the new style of planning serves values beyond efficiency. The U.S. long ago had assumed certain responsibilities for international order. In the judgment of the President, these required military reaction to Communist aggression in South Vietnam. The question was: how much reaction? Considerations of morality and international politics demanded that we should not escalate our military response beyond the level needed to deal with the threat. But limited war has proved a difficult policy for modern democracies, including the U.S., to pursue. Inevitably, voices are raised calling for a much larger margin of superiority over the enemy. Expectable also is the clamor from a group that wants a smaller effort – or none at all. Unless the President had confidence that his Defense Department could maintain an adequate, though limited, level of military action he would be impelled toward an all-or-none choice on the magnitude of the war, a position that would be undesirable on grounds both of morality and of international politics.

That the U.S. still retains its limited-war option is due in no small measure to the confidence that a large part of the public and the White House have in McNamara's system of planning; it is deemed sensitive enough and effective enough to maintain a level of operations in Vietnam calibrated to our limited objectives there, and to changing circumstances. True, there have been shortages of specific materials and underestimations of cost. Nevertheless, the Vietnam war thus far has been the best-calculated military supply effort in twentieth-century U.S. history.

If McNamara's planning system successfully demonstrates in Vietnam that the U.S. can maintain an adequate, though limited, response, then the consequences for foreign policy in the next ten years may be profound. Our enemies have known for a long time that the U.S. had (a) overwhelming capability for total war, and (b) a strong preference for no war at all. Enemy opportunity – and U.S. weakness – lay in the wide area between these two propositions. Doubt that the U.S. could carry out a limited military response has encouraged enemy aggressions (e.g. Korea) and has for many years inhibited the flexibility and range of U.S. policy.

The kind of defense planning now being done requires much sharper definition of war aims than has characterized American policy-making in past conflicts. Just as 'drop-in-the-bucket' comments make for sloppy thinking in domestic policy, so absolutist concepts of 'total victory' and 'unconditional surrender' and 'all-out effort' and 'peace at any price' induce a kind of foreign-policy thinking that can alternate disastrously between the poles of apathetic inaction and apocalyptic commitment. McNamara, his systems analysts, and their computers are not only contributing to the practical effectiveness of U.S. action, but raising the moral level of policy by a more conscious and selective attention to the definition of its aims.

Two Kinds of Planning

To sum up: a society committed to radical and unending change has a deep-seated need, previously almost unknown, to develop a sense that it is able to choose its own path by the light of its own values. Nations dominated by ideologues, confident that their problems have been solved by 'the law of history', can seem to satisfy this need by a coercive central planning structure that, in fact, drains off the possibility of planning from all parts of society except the group at the top. For nearly fifty years the decision-making apparatus of the U.S.S.R., for instance, has run against the grain of human progress because it has not diffused among the people the power to shape their future. Now, the Communists begin to admit that their way of planning has serious deficiencies even in the relatively simple activity of distributing goods and services.

But failures elsewhere could not satisfy the U.S. need for a sense that it was using its science and technology to improve the quality of its own life. Several elements of the U.S. problem have been inherent in the way science itself is organized. Tremendous social complexity follows from the specialization of knowledge. Tremendous social frustration and confusion are byproducts of science's legitimate effort to become objective, value free. What we have needed is a non-coercive system that will bring the appropriate sciences to bear more effectively upon practical problems and at the same time put more scientists into an organized, intellectually coherent contact with the debate over values and goals.

The Way and the Whither

This we are getting. Even the most rigorously mathematical type of systems planning is not, strictly speaking, a science because it is directed toward action, not toward objective knowledge as such. Yet this value-oriented activity of systems planning attracts scientists and earns their respect in a way that older, less organized forms of decision making did not. (Even if one concludes that the decision to drop an A-bomb on Hiroshima was correct, the way that the choice was presented to the decision maker was so terrifyingly sloppy that the subsequent moral trauma of many physicists was understandable.) Moreover, as scientists and other experts are brought together in systematic projects, they are forced into serious communication across the boundaries of their specialties. This serves to integrate the total body of knowledge and it also opens science to the laymen because when scientists from various fields speak to one another about practical problems they use language that any serious citizen can understand. In this partial but vital way the new style of planning 'democratizes' science.

The new style of dealing with the future offers to millions of living Americans an opportunity far more significant than material progress. Since Socrates, at least, Western civilization has respected the examined conclusion, the conscious conclusion, the conscious connection between thought and action, the intentional life. That we are now developing a set of more effective methods for shaping the future represents a fundamental advance along a main line of social and individual evolution. That most of our public and private planning is and will continue to be directed toward material ends should not mislead anybody into supposing that there are no supramaterial elements involved in the process itself. By 1977 the U.S. should understand more clearly that its highest satisfactions are derived from the way we go about forming our choices and organizing our action, a way that stresses persuasion over force and arbitrary authority, a way that extends to more and more men shares of responsibility for the future. By 1977 it may be clearer that we are not just pursuing a material 'more'; that what matters to us is how we formulate our goals and how well we pursue them; that in worldly progress, as in another, the destination is inseparably bound up with the way.

Further Reading

The most useful literary stand-by to systems theory is the *General Systems Yearbook*, published by the Society for General Systems Research, Bedford, Massachusetts. For the years prior to 1964, Young (1964), has provided us with a guide telling which authors have concerned themselves with which concepts and, conversely, which concepts leading authors have concerned themselves with. In using this source the reader does need to bear in mind its tendency to be a-historical, to be faddish, and to regard a mathematical formula as an adequate substitute for a conceptual definition.

W. BUCKLEY, *Sociology and Modern Systems Theory*, Prentice-Hall, 1966.

M. L. CADWALLADER, 'The cybernetic analysis of change in complex social organizations', *American Journal of Sociology*, vol. 65 (1959–60), pp. 154–7.

D. EASTON, *A Systems Analysis of Political Life*, Wiley, 1958.

R. R. GRINKER (ed.), *Toward a Unified Theory of Human Behavior*, Basic Books, 1965.

M. HAIRE, 'Biological models', in M. Haire (ed.), *Modern Organization Theory*, Wiley, 1959.

P. HERBST, 'A theory of simple behavior systems', *Human Relations*, vol. 14 (1961), pp. 71–94.

O. LANGE, *Wholes and Parts: A General Theory of System Behavior*, Pergamon, 1965. (Translated from the Polish by E. Lejsa.)

J. MARCH and H. SIMON, *Organizations*, Wiley, 1958.

M. TODA, 'Design of a fungus-eater: a model of human behavior in an unsophisticated environment', *Behavioral Science*, vol. 7 (1962), pp. 164–83.

G. VICKERS, *Towards a Sociology of Management*, Basic Books, 1967.

W. G. WALTER, *The Living Brain*, Duckworth, 1953; Penguin, 1961.

O. R. YOUNG, 'A survey of general systems theory', *General Systems*, vol. 9 (1964), Society for General Systems Research, pp. 61–80.

Acknowledgements

The readings in this volume are acknowledged from the following sources:

Reading 1 The Commonwealth Fund and Harvard University Press
Reading 2 *Philosophical Review*, J. Feibleman and J. W. Friend
Reading 3 Liveright Publishing Corporation
Reading 4 American Association for the Advancement of Science and L. von Bertalanffy
Reading 5 John Wiley and Sons, Inc.
Reading 6 Chapman and Hall, John Wiley and Sons, Inc., and W. R. Ashby
Reading 7 Society for General Systems Research and *Voprosy Filosofii*
Reading 8 G. Sommerhoff
Reading 9 *British Journal of Psychiatry* and M. P. Schützenberger
Reading 10 American Psychological Association
Reading 11 Chapman and Hall, John Wiley and Sons, Inc., and W. R. Ashby
Reading 12 Plenum Publishing Company Limited
Reading 13 American Sociological Association and P. Selznick
Reading 14 Pergamon Press and the Commonwealth and International Library of Science, Technology, Engineering and Liberal Studies
Reading 15 Free Press of Glencoe
Reading 16 University of Pennsylvania (Management Science Center) and Society for General Systems Research
Reading 17 *Behavioral Science*, A. O. Hirschman and C. E. Lindblom
Reading 18 *Fortune*

Author Index

Subject Index

Subject Index

Penguin Modern Management Readings

Business Strategy
Edited by Igor Ansoff

Collective Bargaining
Edited by Allan Flanders

Consumer Behaviour
Edited by A. S. C. Ehrenberg and F. G. Pyatt

Design of Jobs
Edited by Louis E. Davis and James C. Taylor

Management and Motivation
Edited by Victor H. Vroom and Edward L. Deci

Management and the Social Sciences
Edited by T. Lupton

Management Decision Making
Edited by Lawrence A. Cyert and Richard M. Welsch

Management of Change and Conflict
Edited by John M. Thomas and Warren G. Bennis

Management of Production
Edited by M. K. Starr

Marketing Research
Edited by Joseph Seibert and Gordon Wills

Modern Financial Management
Edited by B. V. Carsberg and H. C. Edey

Modern Marketing Management
Edited by R. J. Lawrence and M. J. Thomas

Organization Theory
Edited by D. S. Pugh

Organizational Growth and Development
Edited by W. H. Starbuck

Payment Systems
Edited by T. Lupton

Programming for Optimal Decisions
Edited by P. H. Moore and S. D. Hodges

Systems Analysis
Edited by S. L. Optner

Trade Unions
Edited by W. E. J. McCarthy